Sensing Changes

The Nature | History | Society series is devoted to the publication of high-quality scholarship in environmental history and allied fields. Its broad compass is signalled by its title: nature because it takes the natural world seriously; history because it aims to foster work that has temporal depth; and society because its essential concern is with the interface between nature and society, broadly conceived. The series is avowedly interdisciplinary and is open to the work of anthropologists, ecologists, historians, geographers, literary scholars, political scientists, sociologists, and others whose interests resonate with its mandate. It offers a timely outlet for lively, innovative, and well-written work on the interaction of people and nature through time in North America.

General Editor: Graeme Wynn, University of British Columbia

NATURE | HISTORY | SOCIETY

Sensing Changes

Technologies, Environments, and the Everyday, 1953-2003

JOY PARR

http://megaprojects.uwo.ca
with new media by Jon van der Veen

FOREWORD BY GRAEME WYNN

UBC Press • Vancouver • Toronto
University of Washington Press • Seattle

20 19 18 17 16 15 14 13 12 11 10   5 4 3 2 1

Printed in Canada on FSC-certified ancient-forest-free paper (100% post-consumer recycled) that is processed chlorine- and acid-free.

Library and Archives Canada Cataloguing in Publication

Parr, Joy, 1949-
   Sensing changes : technologies, environments, and the everyday, 1953-2003 / Joy Parr ; foreword by Graeme Wynn.

"http://megaprojects.uwo.ca, with new media by Jon van der Veen".
Includes bibliographical references and index.
ISBN 978-0-7748-1723-3 (bound); 978-0-7748-1724-0 (pbk); 978-0-7748-1725-7 (e-book)

   1. Economic development projects – Environmental aspects – Canada.
2. Economic development projects – Social aspects – Canada. 3. Ecological disturbances – Canada. 4. Human ecology – Canada. 5. Human beings – Effect of environment on – Canada. 6. Nature – Effect of human beings on – Canada.
7. Human ecology – History. 8. Traditional ecological knowledge. I. Title.

HM856.P37 2010   304.20971   C2009-906370-0

Canadä

UBC Press gratefully acknowledges the financial support for its publishing program provided by the Government of Canada (through the Canada Book Fund), the Canada Council for the Arts, and the British Columbia Arts Council. This book has been published with the help of a grant from the Canadian Federation for the Humanities and Social Sciences, through the Aid to Scholarly Publications Pro-gramme, using funds provided by the Social Sciences and Humanities Research Council of Canada. The author acknowledges the assistance of the J.B. Smallman Publication Fund, and the Faculty of Social Science, The University of Western Ontario.

UBC Press
The University of British Columbia
2029 West Mall
Vancouver, BC V6T 1Z2
www.ubcpress.ca

University of Washington Press
PO Box 50096
Seattle, WA
98145-5096, USA
www.washington.edu/uwpress

In grateful memory of

ATHNASIOS (TOM) ASIMAKOPULOS (1930-90),

McGill University economist,

enthusiast,

whose teaching of theory made us attend,

first and last,

to the assumptions

through which we simplified the material world.

Contents

Illustrations

TABLES

"Now I am Ready to Tell How Bodies Are Changed Into Different Bodies"

Graeme Wynn

In 1933, the London firm of Chatto and Windus published *Culture and Environment* by Cambridge don and literary critic F.R. Leavis and poet Denys Thompson. Remarkably, this little book, addressed to school teachers, was intended to ensure the survival of English life and literature. Ultimately, it reflected early interwar concerns about the increase in leisure time and consumerism, anxieties over the meaningless nature of modern work, fretfulness about the general public's incapacity to engage in positive forms of recreation, disquiet at rising levels of unemployment after 1929, and a critique of industrialism voiced by Matthew Arnold in the nineteenth century. On the face of it, however, *Culture and Environment* was an attack on the enemies of good taste and sensibility – "films, newspapers, advertising – indeed, the whole world outside the class-room" – and an assertion that "it is on literary tradition that the office of maintaining continuity must rest."[1]

Leavis and Thompson accepted that literature was only a substitute for experience, but in the circumstances of the times they regarded it as the best hope for the future. Their book opens with a lament for the loss of an earlier England, for "the organic community with the living culture it embodied." In their view, "folk-songs, folk-dances, Cotswold cottages, and handicraft products" were "signs and expressions of something more: an art of life, a way of living, ordered and patterned, involving social arts, codes of intercourse and a responsive adjustment, growing out of immemorial experience, to the natural environment and the rhythm of the year." There was a time, in other words, when "English people did once have a culture" but it had been destroyed by the machine and all that it brought with it in the way of standardization and levelling-down. This, said Leavis and Thompson, had produced a loss of oral tradition, and of "the memory that preserves the 'picked experience of ages,'" since it had debased words, emotional life, and the quality of living. There could be no going back, but literature could substitute for talk, foster the remembrance of things past, and sustain the sense of "something more."

There is a mythic quality to this history, and the logic of the argument it seeks to advance is weak. So Raymond Williams has pointed out that this picture of olde England ignores the "the penury, the petty tyranny, the disease and mortality, the ignorance and frustrated intelligence" that marked earlier times. Likewise, literary scholars have wondered how to square the circle in the argument that literature maintains tradition because "the vigour and potency of words" depends upon their association with tradition when the tradition to be maintained is dead.[2] But the book's ideas had consequences, not least for a young Manitoban studying at Cambridge in the mid-1930s. Marshall McLuhan soon fell under the influence of F.R. Leavis and his wife, Q.D. Leavis. From them he came to understand that "practical literary criticism could be associated with training in awareness of the environment," and to see writing as a response to different reading publics. *Culture and Environment,* concludes Marshall McLuhan's biographer Philip Marchand, "helped nudge ... [him] away from being a purely literary critic ... to becoming a student of society and eventually the media."[3]

From these beginnings McLuhan shaped an intellectual position that made him a world-leading figure in the study of popular culture and communications media. A critical element of McLuhan's developing thought lay in the distinction he drew between visual and acoustic space. On this account, visual space is a concept, a construct of the eye, a space demarcated with vision. It is typically regular and static, it encourages a linear way of thinking, and it has long been the dominant mode by which educated members of Western civilization have apprehended the world. Acoustic space, by contrast, is a percept – unenclosed, dynamic, and interactive. It lacks a primary centre and is distinguished by the inclusivity and simultaneity, the "allatonceness" (to borrow McLuhan's term), of its resonances.[4] The distinction between concept and percept is important here, and helps clarify the fundamental point at issue. For McLuhan, concepts are detached systems that neutralize participation by explaining the world; they distance people from objects by encouraging passivity and producing inventories. Percepts, by contrast, are participatory and engaged; they elicit feelings and draw upon the full range of senses – smell, sound, taste, and touch, as well as sight. "Perception is mercurial ... instantaneous, boundless, and involving. Conceptualization is static, repetitive, detached, and self-enveloping."[5]

All of this warrants reprise because it provides a valuable platform for the contextualization and deeper appreciation of *Sensing Changes*, Joy Parr's perceptive exploration of local peoples' embodied understandings of the mixed set of environmental and technological changes that affected everyday lives in different parts of Canada after 1945. Like Parr's previous books, *Labouring Children*, *The Gender of Breadwinners*, and *Domestic Goods*, *Sensing Changes* challenges readers to think again about the frequently taken-for-granted.[6] How, she asks, do people know their worlds? And how do changes in those worlds affect their being? Adopting a firmly materialist perspective, and maintaining a resolute focus on technologies, environments, and everyday practices, she argues that experience – physical, embodied, tactile, textured, sensuous experience – precedes the symbolic construction of meaning through thought or language in the process of understanding the settings and circumstances of human lives. In other words, Parr urges us to recognize that people make sense of the world "directly through their sensing bodies" and that "human bodies are conditioned by the circumstances of time and place" (8). In these pages, bodies are multifaceted instruments registering the sights, sounds, smells, touches, and tastes through which people develop their awareness of the world, they are storehouses of phenomenological knowledge about places, and they are laboratories for experiment and adaptation in reaction to change. In bringing us to our senses, so to speak, this book offers a fresh perspective on the past.

Writing early in the twenty-first century, Parr self-effacingly suggests that this perspective should be easy enough for historians and geographers to accept, because they are engaged in "fundamentally contextualizing crafts, committed centrally to recuperation of elements once common sense, relationships enduring in the place and now passed out of mind" (8). But I am less certain, and less sanguine, especially in considering these ideas against the canvas of past practice in these disciplines. Both, it seems to me, have strong concept-ual pedigrees built upon their commitment to the assumptions and structures of visual space and their embrace of what the historian of sound Mark M. Smith once called "an Enlightenment conceit with visuality."[7] In practice if not by explicit precept, generations of historical scholars honoured the conviction, expressed by Joseph

Addison in the English magazine the *Spectator* early in the eighteenth century, that "OUR sight is the most perfect and most delightful of all our senses."[8] Into the late twentieth century at least, most historians seemed comfortable portraying bygone times through the eyes of historical actors (whose apprehension of events was, of course, generally retrieved from written or, in McLuhan's terms, highly visual records), and drawing upon the works of photographers as "illustrations" to offer "an unwittingly visualist representation of the past."[9] Indeed, one might still harbour the suspicion that initiating a conversation about a study of the senses is quite likely to be heard and understood by colleagues as a proposal to study the census.

Geographers, with their emphasis on space, place, and the look of the land have been as resolutely committed as any group of scholars over the years to working within the visual space paradigm. Much of the discipline is devoted to "visualizing," whether through maps, diagrams, charts, and other image forms, through fieldwork intended to develop "an eye for the country," or through the practice of geographical synthesis or seeing together the constituent elements of a scene or landscape. Indeed, Denis Cosgrove authored a series of elegant and cogent contributions associating geography's visual bias with the discipline's embrace of the landscape idea, equating the geographers' use of the concept of landscape with Western traditions of landscape painting, and arguing that this way of seeing worked to entrench notions of objective knowledge and visual authority.[10] According to one of its practitioners, who perhaps chafed a little at this realization almost thirty years ago, "Geography is to such an extent a visual discipline that, unique among the social sciences, sight is almost a prerequisite for its pursuit."[11]

Now back to McLuhan, who understood as well as anyone that things are not fixed. One import of his well-known formulation, "the medium is the message," is that each medium "impose[s] its own spatial assumptions and structures" on consumers. From this it follows, as Richard Cavell has noted in his study of McLuhan as a spatial theorist, that "visual space was only one kind of space, and [that] as electronic media brought the other senses back into analogical interrelationship, other sorts of spaces would come (back) into being, spaces that would be dynamic and interactive."[12] In recent years, new media – the Internet, search engines, social networking sites, and, more darkly, various forms of digital surveillance – have produced a tsunami of change in intellectual and everyday life. We now live in something rather akin to acoustic space, in a world that is increasingly global, inclusive, and synchronic, and this has (as McLuhan anticipated) begun to transform world views, scholarly practices, and customary patterns of behaviour.

The term "multitasking" has jumped from the black box of our computers (where it describes the process of context switching, or rapidly reassigning a CPU [central processing unit] from one task to another) to daily life (where it marks the allocation of CPA [continuous partial attention] to a rapidly rising tide of information and responsibilities). Classrooms (and learning and knowing) have been transformed as rigid rows of seats have given way to flexible configurations; "sages-on-stages" have been superseded by "guides-at-the-side"; dispersed, student-centred learning has replaced the linear transmission of information; podcasts available anywhere, anytime threaten the static (fixed in time and space) gathering "in class"; and service learning is embraced, in part, because it replaces the linear inflexibility of traditional teaching with dynamic, collaborative, communal engagement.

Academic practice, scholarship in the disciplines, has also felt the impact of these changes. Indeed, the very traditional idea of disciplines (demarcated, defined, fixed) is weakening as interdisciplinarity has become a watchword and boundary-crossing a norm. In recent years, many geographers, influenced by deconstructive impulses to question taken-for-granted assumptions and to ask how and why particular interpretations become accepted, have grown critical of the Cartesian (or visual) emphases of their field (in terms McLuhan would have understood) as linear, mechanistic, nationalistic, and fixed and have sought a more multifaceted, inter-relational, discursive understanding of the contemporary world.[13] History, too, has responded to these influences as practitioners have redefined their aims, probed new dimensions of the past, adapted new research methodologies, and experimented with new forms of presentation reflecting the concerns of our times. But however moved they may be by their existence in acoustic space, historians studying what are, in the main, visual-space times have generally found it difficult to bridge and reconcile the two worlds.

Of course things are neither as neat nor as simple as this brief sketch suggests. Thought-provoking as McLuhan's division of the past into visual and acoustic spaces may be, it is ultimately as vulnerable to qualification and question as Leavis and Thompson's "then and now" portrayal of English history. Both depend upon mythopoesis, the invention of a narrative to explain how the world and humankind reached their current state. And both necessarily fail to acknowledge that the enormous and varied *complexities of our naughty world constantly wriggle beyond, and escape, our efforts to wrestle them into neat categories.*[14] Precursors and reactionaries, prophets and visionaries complicate straightforward narratives to the point of undermining their coherence. Hidden riffles, and little islands of thought, create eddies and swirling currents in the braided and ever-contingent stream of time. All of this is to say that tempting (and not entirely inaccurate) though it may be to associate *Sensing Changes* with the inception of acoustic space (because it offers a more holistic view of the past than much earlier scholarship, hybridizes "oral and literate modalities," and articulates what McLuhan called an "integral awareness"), to do only this is to short-change both Joy Parr's achievement and the contributions of other scholars whose influences are registered in the pages that follow.

A few years ago, the appearance of a new journal announced a "sensorial revolution in the humanities and social sciences," but however dramatic this upheaval, its influence upon the writing of history in North America remains limited.[15] Early in the 1990s, sixteen textbooks treating twentieth-century US history were found wanting in any "analytically significant" discussion of the "smells, tastes, sights, sounds, and tactile sensations" of the past.[16] Fifteen years on, a survey of Canadian texts would yield no different conclusion. From the spate of historical research monographs that have appeared in the last decade or so, one can draw a slightly fuller shelf of works engaging the senses, and environmental historians have begun to examine what Linda Nash calls the "inescapable ecologies" imbricated in the relations between human bodies and the world beyond their skin.[17] In Canada, however, such pickings are still thin.

Against this backdrop, *Sensing Changes* stands as a pioneering contribution to Canadian historical writing. Through half-a-dozen carefully chosen, closely textured case studies, Parr reveals the "gritty specificity" of history as she unfolds an argument for the importance of understanding how bodies adapt(ed) to changing times, places, and practices. Each of her local narratives centres on peoples' responses to relatively dramatic alterations in their natural and/or built environments and on the ways in which these changes produce a cognitive and cultural reordering of the ways in which they understood and behaved in place. Together these accounts demonstrate how the patterns and practices of everyday life form an intricate tapestry of threads drawn from the social, technological, environmental, psychological, aesthetic, and other realms that regrettably remain all too separate one from another in the work of scholars trained to comprehend the world from within the particular perspectives of a particular discipline.

The robustness of these distinctions is often surprising. Even within single fields of endeavour, practitioners with different interests seem to march guardedly side by side more often than they appear to walk gleefully arm in arm. Journals mark out sub-fields and contributors use their pages to talk to others who self-identify as fellow travellers. Occasionally, authors from one realm enter another to lament the lack of more active engagement between them. So social historians, historians of science, and historians of technology have danced wary courtships with environmental historians (and vice versa), but the various houses have yet to be formally united. Still, at some slight remove (and with seemingly unavoidable recourse to a metaphor forged in visual space), the separate fields of endeavour merge into a broader landscape, the vitality of which rests upon the effective integration of its constituent elements.[18]

In this long view, it is clear that *Sensing Changes* has varied and interesting antecedents (some of them in parts of the landscape far from the section labelled "history") and a good deal to contribute to the understanding of modern Canada. Anthropologists, who have by and large led work on the senses, at least in the social sciences and humanities, sometimes trace the beginnings of their interest in the topic to Mary Douglas whose *Purity and Danger,* published in 1966, argued that the body and its parts "afford a source of symbols for other complex structures." Following this lead, others explored the symbolic roles of bodily functions and forms across cultures, but as David Howes, director of the Concordia University Sensoria Research Team in Montreal and the general editor of the Sensory Formations book series, has noted, "most of this work ... was curiously desensualized." Only since the early 1990s has the anthropology of the senses become a significant and discrete field of scholarly endeavour.[19]

Historians can push the roots of their discipline's interest in the senses even further back in time. In 1947 the *Annales* historian Lucien Febvre devoted a section in *Le problème de l'incroyance au XVIe siècle. La religion de Rabelais* to a discussion of "Odeurs, saveurs et sons" (Smells, Tastes, and Sounds), and a decade and a half later Robert Mandrou included a discussion of the senses, including a history of hearing loss, in his *Introduction à la France Moderne: Essai de psychologie historique 1500-1640.*[20] In a similar vein, Annaliste Jacques Le Goff, writing in 1960, distinguished church time from merchants' time in medieval Europe.[21] More recently, Alain Corbin extended these beginnings with rich and finely shaded studies of smells and sounds in the French countryside.[22] Most of this work had little influence on scholarship beyond France until it was translated in the last quarter of the twentieth century. For those working in English,

E.P. Thompson's landmark 1967 account of the ways in which those caught up in the web of early industrialization developed new forms of time-consciousness and offered a persuasive (but not initially widely accepted) invitation to think more about "how individuals experienced, understood, made sense of, and invented their environments and themselves" by demonstrating how time was understood bodily, how it was communicated through sound as well as sight, and how it was reconfigured by precepts intended to alter familiar practice.[23] These works stand as precursors to *Sensing Changes* but, rooted in anthropology and social and cultural history, none goes as far as this book in linking the theoretical literatures on embodiment with the concerns of environmental history and science and technology studies.

What, then, does Parr's work add to our understanding of modern Canada and ourselves? First, by recounting the metamorphoses of people and places across the country (and thus giving substance to the opening lines of Ted Hughes' translation of *Tales from Ovid*: "Now I am ready to tell how bodies are changed / Into different bodies"),[24] *Sensing Changes* forces readers to recognize that lives skinned of memory are lives skinned of meaning.[25] Time and again, Parr's case studies demonstrate how transformations in the settings and practices of everyday life alter the ways in which people know themselves, their place, and their work. Ready though we are to accept that traditional knowledge is local knowledge and that relocating those who depend upon such understanding is, in effect, asking them to live their lives in a radically foreign tongue (with all of the stumbling and inaccuracy and incomprehension implied by that), we often forget the vital reassuring benefits of the place attachment, or familiar comfort with setting, that many people develop in relatively short order. As the waters rose behind the Long Sault Dam on the St. Lawrence Seaway in early July 1958, one resident of the area turned to his tearful wife and said simply, tenderly and affectingly: "There goes our youth."[26] Much is lost when the past is shaved from the landscape or when familiarity is filed away by change or displacement. By acknowledging this, Parr reminds us of the price paid by those left lamenting losses, looking back in anger, reminiscing with longing, or building up the "sensory calluses" required to move on: in doing so, she reaffirms the profoundly humanistic warrant of historical scholarship.

A second recurrent, if implicit, theme here is that modern-day Canadians have rarely counted the full costs of their society's enormous technological capacity to alter the environment, or of its collective commitment to industrializing nature in the name of progress.[27] As in the Arrow Lakes

district of British Columbia (the subject of Chapter 5), those whose accidental plight placed them in the path of development schemes have often
found themselves dismissed as "people in the way" of important projects
and their interests discounted as impediments to growth or obstacles
preventing the satisfaction of larger, pressing societal needs. This story
assumes particular poignancy when state-driven megaprojects force the
dislocation of people from their home-places because such actions raise
questions (as the title of Chapter 2 suggests) about place and citizenship.
But other imperatives, operating at different scales, provide many heart-
rending counterparts to tales of displacement by officially orchestrated
development schemes.

Environmental as well as human costs have typically been missing from
or discounted in the "balance sheets" of these various projects, and the
third contribution rendered by Parr's integration of disciplinary perspectives in *Sensing Changes* is an encouragement to think anew about discussions of sustainability and resilience in Canada and elsewhere. To
date, much of this discussion has focused on the economic and environmental dimensions of sustainability; social sustainability, the third leg of
the sustainable development stool defined by the Brundtland Commission
in 1987, has proven more difficult to grasp. But Parr's discussions of the
effects of displacement on individuals and communities are an invitation
to think harder about human resilience and social sustainability. By standard economic and environmental measures, the settlements of Summer
Hill and Lawfield Road, evacuated to establish the Gagetown military
base, were settlements of poor people on poor lands. Yet these settlements
were rich in shared experience, community leadership, and mutual aid,
and those who lived in these places had a sense of belonging that sustained
them, and made them resilient against long odds – at least until they no
longer had access to the communities to which they belonged.

Parr also gives us an arresting clarion call from the other side of the
country from the pen of Donald Waterfield, author and farmer along
the Columbia as it was turned from river to reservoir. Commenting on
the 1956 Report of the Royal Commission on Canada's Economic Prospects,
he asked, "Have we become so confused by the complexity of modern
economics that we can no longer distinguish between good and bad?"
Were Canadians blind to the "'aesthetic value' of working to live rather
than working to produce, when 'beyond the acquisition of bare necessities,
we all have food, clothes and shelter, and most of us, entertainment as
well'"? Here and in other writings, Waterfield called into question the
view of development described by the environmental historian Donald

Worster "as a single cultural standard against which all people could be measured."[28] It was Waterfield's great insight to counsel moderation rather than "unlimited materialism and accumulation" and to advocate stewardship rather than exploitation of nature (118). Fifty some years on, Waterfield's questions remain pertinent and his intellectual position worth contemplation. Amid rising debate about expanding ecological footprints, declining fossil fuel supplies, and the wisdom of eating locally produced food, it is good (and humbling) to remember that we are not the first to walk this line of argument.

Fourth, let it be said that *Sensing Changes* is a beginning rather than an end and a methodological as well as a substantive contribution to the Canadian literature. By asking new historical questions, it illuminates important and neglected aspects of the past. Other facets of the story that Parr has outlined beg investigation, and their interrogation will raise new questions about the ways in which history is practised. Because here it must be noted that for all its topical and scholarly freshness, the book you hold in your hand (shaped by the constraints of print as a medium) is a methodologically transitional work struggling (as McLuhan might have said) to reconcile its roots in visual space with the concerns of our increasingly acoustic space times. With Jon van der Veen, Parr has sought to address this tension by developing new media explorations of most of the settings discussed in this book (available at: http://megaprojects.uwo.ca). There, oral recordings, air photographs, documents, and other materials are brought together to amplify and extend the limitations of the printed page and allow those who follow the links to come a little closer to experiencing the sites of investigation. But questions remain: Can we ever hope to hear and smell and touch and taste the past? Can we apprehend bygone eras and lost places as they were sensed by those who lived in them? Indeed, in what sense can we claim (for all of our visualist representations of it) to see the past?

Although *Sensing Changes* focuses for the most part on megaprojects of one sort or another (the building of big dams and nuclear generating facilities, the excavation of ship canals, and the establishment of a military training ground), and its final chapter examines a tragic and infamous case of failure by a public utility to ensure the supply of untainted water to Walkerton, Ontario, it would be a mistake to conclude, from these pages, that only such dramatic and disruptive events affect peoples' phenomenological ways of knowing or, indeed, that changes in embodied practice are always negative or traumatic. Parr's aim is to revise and modulate our understanding of the past by drawing attention to the ways in which environmental, technological, and social changes are felt in ways that people know but cannot always tell. Her mandate is not to catalogue all the ways and instances in which knowledge held "beneath the skin" is altered by changing material circumstances (10). Her case studies make her case, and they do so in dramatic fashion. Still, readers must not make of megaprojects the sort of deus ex machina that Leavis and Thompson conjured from industrialization in *Culture and Environment.* Change in one form or another is a staple of history, and people have responded to change intellectually and bodily in countless different ways. They have moved and changed of their own volition as well as under the forced and unanticipated circumstances detailed in these pages. Loss and dislocation have been balanced by hope and adaptation, and the sharp edges of sensed changes have been blunted by mindful visions of new things to come.

None of this is to minimize the sense of loss, the emotional toll, or the sensory assault produced by the silencing of river rapids, the inundation of familiar acres, the waft of the rotten egg smell of hydrogen sulphide across a popular campground, or the erosion of trust occasioned by good water turning bad. But just as "folk-songs, folk-dances, Cotswold cottages, and handicraft products," were not the sum of "olde England," so these instances are but part of the sensory history of Canada. Consider by way of illustration what I take to be one of the more poignant observations in a poignant story about the losses occasioned by the establishment of Camp Gagetown: Eugene Morris' recalled "lost delight in things that have grown under our moulding" (38). And compare this with the affecting lines penned by Charles Bruce, Nova Scotia newspaperman, author, and poet, chronicler of the slow changes that reshaped his imaginary Channel Shore between the great wars of the twentieth century, who wrote, in "Orchard in the Woods," of a place where:

> Red spruce and fir have crossed the broken lines
> Where ragged fences ran ...

In this place "Where oats and timothy" once "moved like leaning water" in the late summer wind, now only "stunted alders and tall ferns" grow. Nothing stands atop the collapsing cellar and a woodchuck burrows beneath "the wreckage of the vanished barn."

> Only the apples trees recall the dream
> That flowered here – in love and sweat and growth,
> Anger and longing.

But in this place, where "Clearing and field and buildings [have] gone to waste" – a hunter heading home in the fall

> will halt a moment, lift a hand to reach
> One dusky branch above the crooked track,
> And, thinking idly of his kitchen fire,
> Bite to the small black shining seeds and learn
> The taste of ninety seasons, hard and sweet.[29]

This small example from eastern Canada, and Parr's six larger case studies from across the country suggest, as the environmental historian Roderick Frazier Nash once observed, that all places are "multi-faceted composite[s] of history, natural environment, and personal experience." In demonstrating this, Parr has moved us to a deeper understanding of the process of place making. By reminding us of the importance of the senses through which people engage their settings and constitute their experience, she has also thrown down a gauntlet for future scholarship: to pay close attention to the ways in which people sense and value places. Doing so, one might hope, could help unlock "the secret to living lovingly and sustainably on this endangered planet."[30]

The Megaprojects New Media Series

Jon van der Veen

Facing each chapter in this book are introductions to the new media projects associated with that chapter's site of study. The entire series, which we call the Megaprojects New Media series, is available at http:// megaprojects.uwo.ca. Unlike web resources that serve to echo the findings of a book to an online audience or to provide a residual basin for items that could not be included in the book, the Megaprojects New Media series was conceived of as a parallel path of experimentation and scholarship into new media representations of the accounts of people living amid megaprojects.

Design of the Megaprojects New Media series is informed by my doctoral dissertation research at Concordia University into the culture of the database. In contrast to the notion that web-based media may portend the eclipse of the dominance of the printed word, new media forms are in fact often subtended by databases that follow a textual logic of keyword markers, and this approach has been increasingly applied to the organization of video, audio, and image materials.[1] Even the most forward-looking and intriguing approaches to integrating historical scholarship with new-media practices are modelled after such textually dominated database services as Google search and Wikipedia.[2]

However, the Megaprojects New Media series emphasizes the sensual over the searchable. By keeping closer to the visual and auditory material contexts through which we came to know the accounts of those who were affected by the megaprojects discussed in this book, our new media experiments aim to effect a more coherent, sensuous, and memorable reclamation of experience than is possible through textual representations.[3]

Rather than approach our evidentiary documents as "entries" to be tagged and granularly re-contextualized into search results, the materials and their accompanying contexts matter: the hesitations registered in the audiotaped voices from Bruce County are not apparent in the transcriptions; the writing Val Morton etched over his old photographs could not be captured by rendering their content in type; the landscape changes around Base Gagetown are uniquely apparent through diachronic image sets; the Iroquois images and the recollections of its citizens are installed as they were remembered, along the street.[4]

These new media projects are available as resources for users to differently experience the sites shared in this book, and they are tentative, practice-based probes towards sensuous new media forms not organized around keywords and search queries.

Acknowledgments

This project began in conversations about theory with David Howes and John Staudenmaier sj. The empirical research was possible because thoughtful and busy residents of six places decided to donate time and attention to it and to me. They worked their occupational, professional, and personal networks to make my goals and needs known, and lent their credibility to the project. Because these inquiries are about challenges and adaptations that elude texts, without their guided walks and talk; the reflections and corrections, which comprised their mode of teaching; the photos and cassette tapes; the mementos, confidence, and confidences they shared; and the offers of that "one more drive out together to look," there would have been no book. Sometime along the way, matters turned around, and the fact that they wanted and expected a book became my reason to continue.

Travelling from west to east, in the Arrow Lakes of British Columbia I am grateful to Rosemarie and Milton Parent of the Arrow Lakes Historical Society; Colin Preston of CBC Vancouver; the late Stan Rowe, author, jazzman, biologist, and environmental ethicist; and to Janet Spicer, market gardener, the person who provided Stan's household with carrots and me with an introduction to the soils of the Lakes. The late Val Morton came with his dog Patch from Big Eddy Ranch to meet me at the library in Prince George; and Ted Harrison, then a student at the University of Northern British Columbia, made trips out to Big Eddy to create an oral record of Val's chronology of the loss of his land. Val's sister-in-law, Michele Morton, has helped us since. Helen and Oliver Buerge, Charlie

Berry, Ernie and Olive Roberts, Earl Moffatt, Ernie Orr, Glen Olson, Hank Scown, Marlene and Jack Allard, John Brown, Lloyd Parkyn, Lily Grimmett, Ruth and Nigel Waterfield, Olive Robins, Pete Coates, Peter Ewart, Roy Bateman, Thomas Fulkco, Winnie Ehl, and William Barrow spoke with me, reviewed the transcripts, and in many ways thereafter remained connected with the project. In Castlegar and Nelson, Alisha Gray, then a student of historian Duff Sutherland at Selkirk College, read the local newspapers and refitted my categories when they proved insufficiently attentive to Kootenay conditions. Marilyn James, Aboriginal advisor at the college, walked me to Lakes sites, and Eileen Delehanty Pearkes, and Walter Volovsek shared their research about and their sense of the place.

Along the Huron Shore in Ontario between Southampton and Kincardine, my guides were first the newspaper editors, Eric Howald of the *Kincardine Independent*, and Barbara Fisher and Marie Wilson, then of the *Kincardine News*. In Tiverton, the village nearest the Bruce Nuclear Site, the heavy water plants, and Inverhuron Park, I learned much from Eugene Bourgois, shepherd and knitter, and Jim Dalton, plantsman and retired Lummis heavy water project engineer. I am also grateful to Donald and Doris Milne, Dick Joyce, Dave McKee and Frank Caiger-Watson, Barry Schell, Stephen Bell, Jake Hunter, Vern and Fran Austman, Bob Ivings, Charlie Mann, Ben Cleary, Keith and Lola Davidson, Ken Elston, Frank Ruddock, Jim Bayes, James Weir, Jim Dalton, John MacKenzie, Robert Mackenzie, William Mackenzie, Lorne McConnell, Tony McQuail, Paul Carroll, Robert Wilson, and John Wilkinson, many of whom now live some distance away, for their narratives and analysis of the Bruce Site when it was part of a public utility. In these years, I savoured the teasing laughter of an adult friendship with my aunt Margaret Munro, now gone, who raised her girls on the North Line during the years of Douglas Point.

In Walkerton, during the O'Connor Commission hearings, Gerdie Blake, from the Hanover office of the Ontario Federation of Agriculture, set an example of assiduous and informed listening, and I followed along in her slipstream. Later, Bruce Davidson, Charlie Bagnato, and Ron Leavoy of the Concerned Walkerton Citizens answered my questions. John McPhee, editor of the *Walkerton Herald-Times*, area journalist Pat Halpin, and Justin Kraemer, Bruce County GIS specialist, helped me with images and sources. Once I became associated with the WEL investigators, the long-term health study team at the University of Western Ontario, Bill Clark, Marina Salvatore, and Arlene Richard, became colleagues, and I have savoured our times together, both of work and relief from work.

In the "Seaway Valley," the late Fran Laflamme was my publicist and promoter.[1] There could be none better for this role. I expect I am only the last of many who grew through her generosity and forthright guidance. Later, my contact in the eastern end of the valley was Jim Brownell, MLA, Fran's successor as leader of the Lost Villages Historical Society. Dave Dobbie and his staff at Upper Canada Village guided me though their archives and have since accommodated our aspirations for the virtual walking tour through Iroquois. The move from consulting the texts to visiting and taping along Lake St. Lawrence was facilitated by Carleton McGinnis of Morrisburg, who introduced me to Joyce Fader, who introduced me to Iroquois and the work of other heritage activists there. Sandy-Lee Shaver, then editor of the local newspaper, identified for the village the stranger who was eating raspberry pie in the plaza restaurant and steered me away from more than one big mistake. Hilda Banford, Keith Beaupre, Ambert Brown, Gwyneth Casselman, Ray Casselman, Les Cruickshank, Glen Cunningham, Janet Davis, Ron Fader, Jack Fetterly, Shirley Fisher, George Hickey, Shirley Kirkby-Carnegie, Leo Merkley, Caroline Robertson, the late Joseph Roberts, Jean Shaver, Lee Shaver, Isabel Shaw, Rose and the late Frank Sisty, and Carl Van Camp visited me and the first generation Iroquois maps and photos at the Town Hall and shared their personal archives liberally. Eleanor Pietersma, the Iroquois librarian, has been a booster and advocate, as local librarians so often are, and her reports back about how walkers in our "virtual Old Iroquois" have fared have been gratifying and clarifying.[2]

In the Saint John Valley, I am grateful to Margaret Conrad and her household for the bed, board, and conversation they provided during my stays in Fredericton, as well as to the staff at the Provincial Archives of New Brunswick and the Queens County Museum, Gagetown, and the CFB Gagetown Museum and Archives. Around Gagetown I interviewed only Willard Clarke, Maude Underhill, David McKinney, and Connie Denby, so rich was the oral history archive created before me at the New Brunswick Provincial Archives. David McKinney and Connie Denby, his successor as leader of the Base Gagetown Historical Society, have helped when they could, including making space for me at the activities marking the fiftieth anniversary of the displacement from the Baselands back from Gagetown. Connie, a lover of dance, has also helped in the project documenting of the continuing musical evenings at LOL #4 Hall, which my colleagues Greg Marquis and Donn Downes from the University of New Brunswick, Saint John, are leading.

Jan Burnham, a superb teacher of radiation protection theory and practice, has kept a watching brief on the Point Lepreau part of this project, guiding me to sources, lending teaching materials, and drawing me back when I've strayed out of the zone. Through him, I met Brian Patterson, Dan McCaskill, David Meneer, Stephen Frost, Gerald Black, and Ken Hill, pioneers at Lepreau with wide experience in the nuclear industry, men of keen wit and generous spirit.

At the University of Western Ontario, I am especially grateful to Catherine Ross, former dean of the Faculty of Information and Media Studies (FIMS), who prepared the way for my appointment as a Canada Research Chair in Technology, Culture and Risk; and to Joanna Asuncion and John Fracasso, also of FIMS, who accommodated and forwarded this research, even when the needs my team and I articulated seemed at least passing peculiar. Their example is a reminder to me, when my energies wane, that remaining open to the enthusiasms of others usually nourishes more than depletes. My research colleagues in this project at Western have been Denver Nixon, Wendy Daubs, Jessica Van Horssen, and Sandra Lynne Hodgson. Jon van der Veen, now of Concordia University, through creativity, patience, and panache, designed the new media experiments and brought them into being. Fe Alcos, Nancy Doner, and Maggie Nicolson transcribed the tapes and digital files. I am especially grateful to Dan Shrubsole, chair of the Geography Department at Western, for enabling my move into the Faculty of Social Science, and to colleagues there, in these straightened times, for more than merely tolerating two more feet under the dinner table. As you will soon see, the maps, graphs, illustrations, and photographs here are of exceptional quality. These we owe to Trish Connor Reid, the Western geography department cartographer, and her colleague Karen Van Kerkoerle, and to Ian Craig, media and digital imaging specialist in Western's Department of Biology, who turned the much weathered images I had collected into the gems you see here.

At UBC Press, Randy Schmidt, Laraine Coates, and Peter Milroy welcomed the book in all its parts and guided me well. Graeme Wynn's commentaries, particularly in regards to the lower Saint John Valley and the Arrow Lakes, have made this a better book.

My thanks to them all, and most of all to my husband, Greg Levine.

Joy Parr

Sensing Changes

I

INTRODUCTION

Embodied Histories

A s humans, we live in environments, amid technologies, learning by doing. Our bodies are the instruments through which we become aware of the world beyond our skin, the archives in which we store that knowledge and the laboratories in which we retool our senses and practices to changing circumstances. Bodies, in these senses, are historically malleable and contextually specific. Our senses are the conduits through which knowledge of technology and the environment flow and, through retuning habit and reflex, the ways we habituate to our changing habitat.[1] Most of these adaptations are held beyond speech, often outside conscious awareness.

A decade ago I began to study five mid-twentieth-century Canadian megaprojects and along the way became involved with a Canadian community overtaken by a catastrophic water contamination. As I listened to residents in these six contexts, the focus of my attention was drawn from the megaprojects to the processes by which inhabitants adapted to the habitats the megaprojects had transformed. Local people recounted how these radical changes had unsettled their daily lives and forced them to encounter their environment anew and adapt the practices through which to live competently and sustainably day by day. They described a collective relearning of what to infer was ordinary and exceptional in the winds and the weather; in the lakes, rivers, and native vegetation that surrounded them; in the needs of the crops and livestock they tended; and the aquifers, fish, and woods upon which they depended. These were the interfaces where their active sensing bodies, through the technologies they used and the practices they deployed, engaged their environment as habitat – as sites of mutual remaking. Human bodies tuned to one world beyond their skin,[2] which once had seemed "the world," by sensing the differences and, sensing differently, retuned to another. They talked about learning to recognize and work with a different "ordinary" beyond their skin, to find ways to adapt this new "natural" to their purposes. Their accounts of these crises of competence and confidence, when the recognition of these disjunctions led them to revise their inferences from sensuous signs and to reconsider effective practice, changed the course of my journey among them. For here, in these radical disjunctures, were researchable traces of human bodies being made contextually, temporally, and spatially specific. As inhabitants incorporated into their bodies the altered world beyond their skins, awarenesses they usually held beyond telling, as habit and reflex, became urgently speakable. This process of habituation to habitat through the tuning of the senses and the honing of habit and reflex is embodiment. In *Sensing Changes*, we encounter embodiment both as active adaptation to changed circumstances and as "the whispering of ghosts," relics of past successful adaptations to familiar worlds later remade, persisting as familiars, reminders of losses, and also sources of resilience and resources for rebuilding.[3]

This interface between the body and the world has been much theorized,[4] but this theorizing, as Katherine Hayles has noted, has been unchastened by the ornery attrition of instances and is resistant to push-back grounded in the concrete and the corporeal.[5] The balance between universalizing postulates and gritty specificity must be reset if we are to bring research about environments, technologies, and the everyday – turned for a decade towards the different issues of representation, the symbolic, the linguistic, the discursive, and the textual – back towards the materiality and sensuality of direct encounters with the world beyond our skin. By close scrutiny of the environmental and technological change that megaprojects involved and the adaptations among human neighbours these changes required, I aim to capture how the arrival and persistence of the megaprojects remade modes of dwelling and earning a living, the discernment of hazards, and the experience of pleasures at home and at work in time and place. The book you are holding in your hands reports these findings in a conventional textual way. Yet, attempting to communicate through texts, these meanings, awarenesses, and associations, which intrinsically have resisted representation,[6] has been a challenge.

Conventional modes of scholarly publishing are not well suited to this task, nor are many scholars disposed to welcome the implications of the research. Academic traditions are deeply invested in texts and in textual critique as the arbiter of research results.

For this project, these conundrums spawned a parallel path through which to share findings. These new media experiments, a project led by Jon van der Veen of the communications department of Concordia University, aim to convey the embodied histories words alone cannot tell. You may find this Megaprojects New Media series work at http://megaprojects. uwo.ca. Before each new chapter there are introductions to the relevant work available online: the audio and visual presentations, walking tours, chronological archives, geographic information systems reconstructions, music, and videos that are our attempts to more thoroughly reclaim the embodied "lostscapes" residents mourned.

No place is merely local. All of these instances are located within a single national political economy: Canada in the decades after the Second World War. All were implemented along the international boundary between Canada and the United States and, to varying degrees, were responses to the shared North Atlantic political and military climate of the mid-twentieth century. These transnational influences are most apparent in the political influences upon the choice of project sites. Yet the technical demands the succeeding engineering order placed on human residents and on the air, water, soil, fauna, and vegetation upon which they had relied, while locally inflected, reveal challenges that have relevance beyond the particular. While in many nations such large engineering works are now more often questionable and less commonly initiated, their human and broader environmental effects linger in place. And in many parts of the developing world, megaprojects remain essential elements of statecraft. The usefulness of understanding how changing technologies and environments are taken into the habits and practices of sensing bodies, challenging the security of individuals and the viability of their communities, will remain for some time to come.

We begin with cases where the transformations in technologies and landscapes were effected by the large engineering works – megaprojects – common in the twentieth century, and with hinterland residents, those most frequently obliged to cede their habitats to unquestioned priorities of modern statecraft, to power projects, military bases, and transportation infrastructures. The sixth instance, the water contamination, occurred in May 2000 and takes us from the brute force physical disruptions and population displacements of modernist megaprojects to the environmental and technological challenges to embodied knowledge and local sovereignty likely to be more common in the twenty-first century. This instance hinges on a freak weather event, the intensification of livestock production, and the implementation of science-based regulatory regimes to discern environmental hazards. As with the studies of how the neighbours of twentieth-century megaprojects learned how to cope with unbidden technological and environmental change, so the twenty-first century Walkerton case is a Canadian instance with wider resonance and relevance.

Because the historical study of embodiment is relatively new, and the cases and places here will be unfamiliar to many readers, some conceptual and contextual introduction is in order. This chapter makes the case for embodiment, specifically corporeal embodiment, as a historical process. While in some cases language is the filter that organizes perception (this is embodiment as a figure of speech, as it was commonly employed by literary scholars and some historians during and after the linguistic turn of the late twentieth century), often we know directly by sensing and doing, by corporeal embodiment. We make meaning by doing and organise our awareness and skill through bodily practice. The body is a synthesizing instrument that defies the categorical and linear discipline of language and science. The environments and technologies with which we live, play, and work lead us to develop specific modes of bodily attention and perception. This tuned reciprocity among body, environment, and technology has historically allowed humans to feel at home, competent, and safe. Creating environmental histories and histories of technologies, which borrow more prodigiously from sensory history, attending explicitly to these sensuous adaptations and the habits that organize them into practice, captures both the challenges and possibilities posed by changes in habitat and technique.

Chapter 2 concerns the Canadian Forces Base at Gagetown, New Brunswick, which is set back from the Saint John River, near the Atlantic seaboard and just north of the US border. It was established during the Cold War to train NATO troops. The base took up on 958 square kilometres of developed meadows and timber lands created over several generations by Protestant Loyalists, refugees from the American Revolution, and Catholic Irish families fleeing the potato famine. The inhabitants who had made the wood lots and farms ceded lands they had never intended to sell. Thereafter, many were employed on the base, some using Agent

Orange to clear mazes for tanks and training grounds for infantry. The residents' departure was traumatic and, we now know, wrought grave health effects on local nature, both human and non-human. The chapter investigates the making of timber lands and Jersey herds, their unmaking to accommodate the base, and the subsequent remaking of local religious and language identifications when the British Dominion for which residents had been willing to make sacrifices was reconstituted as a bilingual and multicultural Canada. Their music and crafts are chronicles of their displacement. We have also created a series of representations using air photos linked by geographical information systems (GIS) to show the changes in land use and topography that preceded and followed the development of the NATO base.[7]

Two sites – the Bruce Generating Stations on the eastern shore of Lake Huron in Ontario and the Point Lepreau station near the Maine border in New Brunswick – are the locations for Chapter 3. This is a study of how embodiment figured within a nuclear occupational health and safety regime. The workers at these generating stations, mostly men, came to nuclear work from farming, forestry, fisheries, chemical industries, and shipyards – occupations in which the signs of danger were physical. They learned to adapt to work in the radiation fields of nuclear generating stations by reading instruments, attending to proxies for insensible dangers, and modifying their gait and posture to stay within the registering range of their instruments. Canadian radiation protection practices are singular among early nuclear sites in their non-hierarchical work rules. Each employee was responsible for his own safety. Each learned the theory of ionising radiation so as to infer, from the measurements his instruments registered, the bodily practices to keep himself safe. These physical habits and reflexes substituted for the somatic signs of their former workplaces and for the military discipline in nuclear generating stations in the United States and Europe. The elements in the design of Canadian stations, which accommodated these work routines, are part of the story. Another part, less readily apparent in the textual transcripts of the oral histories than audible in the hesitations, repetitions, and intonations of the digital sound records, is the bodily retuning to signs of danger and safety by which men adapted to this technological transformation in their occupation. We have tried to capture some of this sensory information, lost in the transcriptions,[8] through four sound compositions.[9]

Iroquois, one of eight Ontario villages dislocated by the Seaway, is the focus of Chapter 4. First a Loyalist settlement and then a mill village that attracted textile workers from Quebec, Iroquois was the first village to be remade to make way for the St. Lawrence Seaway and hydro-electric projects in 1958. Some wooden houses were moved back the short distance to the new village. Before the inundation, Iroquois' mature hardwood trees were cut down. In the interests of unimpeded navigation, the masonry structures of the village were burned and then reduced to rubble. Those who had lived in brick and stone eighteenth- and nineteenth-century dwellings moved into 1950s bungalows. The three-storey commercial buildings were succeeded by a small plaza, the Victorian churches replaced by 1950s contemporary brick structures. The International Rapids, the sound mark of the old village in all seasons, was silenced, the roaring river replaced by a placid shipping channel behind a series of dams. The much valued fishery ended. This settlement of walking mill workers was remade as a suburb/village, a treeless built environment designed for the car and the commuter. This chapter is about bodies oriented by sound in space and by the scale of built structures. The new media work associated with this chapter consists of three audio walks of the village as it existed before the inundation.[10]

In 1968, the closing of the High Arrow Dam at Castlegar on the Columbia River in British Columbia turned the Arrow Lakes, narrow mountain bodies of flowing water with a nine-metre seasonal variation in levels, into a storage reservoir for the Bonneville Power Authority. The twenty-one-metre difference between full pool and maximum draw down occurred in response to a US utility's demand for electricity. This is the site for Chapter 5. Most valley residents had lived by combining work in the formal and informal economies, logging, ranching, fruit growing, market gardening, dairy, poultry, and sheep-raising. They lived well but were scorned by resource planners of the 1960s as "stump farmers" insufficiently attuned to the "highest and best use" of their home places. The reservoir flooded the rich bottom lands and the transportation infrastructure of the valley. The movement of water, by which residents had judged distance, direction, and depth in their shoreline habitat, changed from a flow to a radical rise and fall, alternately flooding and exposing an arid wasteland. In this industrialized and colonized environment, many described losing their sense of place and their sense of self. Among these were a family of market gardeners, a writer and mixed farmer, and a rancher. The new media work associated with this chapter chronicles several peoples' lives and land with photographs and video and an account of rancher Val Morton's years on the ranch in his own voice and handwriting.[11]

Chapter 6 is a study of the regulatory decision making that closed a beloved Ontario shoreline park to accommodate a large chemical plant, and of local responses to this emerging hazard. The plant emitted hydrogen sulphide, a gas benign at low concentrations, when it can be smelled, and potentially lethal at high concentrations, when it extinguishes the sense of smell by killing olfactory cells. The chapter explores how somatic, scientific, and topographic constraints (a coastal place of strong and seasonally variable winds contained and directed by a nearby steep ancient shoreline) influenced the decision making of campers, cottagers, a pastoral family, and the scientific staff of the federal and provincial regulators. The audio compositions about danger and nuclear work linked to Chapter 4 also include material gathered from pastoralists and cottagers recounting their growing sense of danger amid uncertainty in and near the park.[12]

We move into the twenty-first century with the case discussed in Chapter 7. In Walkerton, Ontario, unprecedentedly heavy spring rains falling in May 2000 on fields underlain by permeable limestone karst carried surface runoff containing *E. coli* 0157 H7 from cattle manure into a shallow unprotected well, contaminating the drinking water of the town. Seven residents, all elders and children, died of the effects of acute gastroenteritis, and half of the four thousand people in Walkerton that holiday weekend became ill. Many have suffered lingering physical health effects. Walkerton was one of many Canadian communities in which popular suspicions of chlorination made municipal evasions of mandated drinking water quality guidelines acceptable and commonplace. Contemporary neoliberal cost-cutting reduced the effectiveness of regimes whose purpose was to protect public health across the country. Events in Walkerton exposed the grave inadequacy of local ways of judging the safety of water; they revealed how climate change might cause long established agricultural practices to become hazardous; and they made apparent the challenges that increasingly dense human and animal populations sharing a watershed posed to the environment and to health. The new media work associated with this chapter draws on the Walkerton Inquiry transcripts to feature the shifting contexts and meanings of selected utterances such as "good water."[13]

The Conclusion revisits the theme of historically specific bodies. Most historical studies of the consequences of large-scale environmental and technological change focus on property losses and effects upon physical health and employment opportunities. This framing leaves "the people in

the way" with "no place to put" the disorganization they experience in
their sensuous judgments, habitats, and workplaces. The grief for which
they could not find words is often dismissed as nostalgia. At the end of
the Conclusion there is an epilogue that brings together accounts of how
many in these sites have dealt with this estrangement by action-oriented
projects.

"Embodied history" is a term Pierre Bourdieu used to invoke the "active
presence of the whole past ... internalised as second nature and so forgotten as history."[14] The idea is not new. In the mid-nineteenth century, Marx
had argued for "the forming of the five senses ... [as] labour of the entire
history of the world down to the present."[15] That human bodies are conditioned by the circumstances of time and place is easy enough for historians and geographers to grasp. Ours are fundamentally contextualizing
crafts, committed centrally to the recuperation of elements that were once
common sense, relationships enduring in the place and now passed out
of mind.

The next step towards embodied histories is not yet commonsensical
to most scholars. Bodies are not only being *conditioned* by circumstances,
they are also enduring reservoirs of past practice, which *actively influence*
subsequent responses. The body can hold what has passed out of mind.
Bourdieu affirms that awareness honed in practice might be "internalised
as second nature" and persist, "palpable and absent,"[16] even when "forgotten as history." These assertions refuse mental habits we hold so reflexively that we have forgotten their presence and can scarcely discern their
influence: the Cartesian division between body and mind; the founding
premise of recent social construction theories, which holds that knowledge
precedes experience; and Foucault's "epistemological view of the body as
existing only in discourse."[17] The studies I present are founded on different epistemological assumptions: that minds are embodied; that doing
can organize knowing; that logic can be founded in practice; and that
knowledge can elude symbolic representation in texts, the central conduits
through which historians in particular have interrogated the past.

In two 1995 essays in environmental history collections,[18] Katherine
Hayles made the case for a more "embodied" historical practice. She wrote
at a time when social scientists were in the thrall of the "linguistic turn,"
exploring how discourse made and held meanings. Amid the enthusiasm
for Foucault, Scott, and Latour, and the celebration of the many useful
insights we might borrow from literary scholars, her contribution was

neither nourished nor taken up widely by others. It is time to try again – not to turn our backs on the representations of tools and places in language but, rather, to open interpretive space in which to study the robust materiality of technologies, environments, and the everyday, to encounter them as directly and as fleshly as possible, rather than as codified symbolically as language. How might we do this, and why?

I used the unusual word "fleshly" to focus attention on the body as a way of knowing. What humans know and how they organize and reason with that knowledge is "marked by the particularities of our circumstances as embodied human creatures."[19] What are these particularities? Some of these we can assume persist over long stretches of history and across cultures.[20] Humans stand upright and are a certain distance above the ground when they crouch or sit to rest. They walk and run within a certain range of speeds and can reach to touch across a certain distance. Within a certain range, they can retain their balance while moving on slopes and shifting ground. These we could group as proprioception, the sense of bodily knowing in space; kinesthetics, the gait, pace, and posture with which the moving body encounters its surroundings; and proxemics,[21] the emotional comfort with nearness and distance. Some of these change with the lifecycle and over time. A child's sense of "too high" is different from an adult's; the medieval sense of "close quarters" in a dwelling was different from yours and mine. Some are altered by contemporary technologies. Think of the difference in "an hour away" to a walker, a cyclist, and an air traveller, or of "clean enough for comfort" in a household with a washing machine rather than a scrub-board, or a vacuum cleaner rather than a broom.[22]

Much of the bodily knowledge that comes from interactions with the world is not readily captured in words. Michael Polanyi called it "tacit knowledge,"[23] what "we know but cannot tell." Pierre Bourdieu called it "habitus," following on the work of Marcel Mauss.[24] Mark Hansen calls it "experiential excess"[25] – excess because, while it is securely held in bodily experience, it eludes expression through language. In 1992, the eminent environmental historian William Cronon made a disheartened attempt to make the "linguistic turn" in his own work.[26] Respectful though he was of his literary colleagues' insights, he found so much tacit knowledge and experiential excess in the world he wanted to know that he concluded that the narrative form was "dangerous" for his purposes. The ecological senses of non-linearity and randomness, which were the focus of his attention, *by their nature* were synthetic rather than categorical. In their way of being, they were fluxes and fusions, which the strict linear progression of words, what he called the "rhetorical razor" of discourse and narrative, could not adequately convey.

These sensuous ways of interacting with the world are best distinguished as phenomenological or corporeal embodiments.[27] By their resistance to communication in words, the parts of technologies, environments, and everyday practices accessible through these senses are those most marginalized by the methodological turn to discourse analysis. Because they are so central to the processes of dwelling and work, it is hard not to agree with Cronon that bringing them into the foreground of environmental history and the history of technology and the everyday should be a priority.

Many of the senses that come readily to the conscious mind – sight, hearing, touch, taste, and smell – in our time are readily expressed in words. Sight and hearing are the ways humans interact most externally with the world and, thus, are the most readily verifiable and amenable to the standards of empirical testing and replication that scientists prefer. Comparatively, it is relatively easy to agree that we are seeing or hearing the same thing as the person beside us. These are also the senses cultural critics and science studies scholars are most likely to borrow as they search for bodily metaphors to express socially constructed institutions of knowledge. Think of how "panopticon" is used to convey the power to monitor in many dimensions, or "resonance" to suggest the diffuse meanings that emerge from a source. This is not corporeal embodiment but, rather, epistemological embodiment or embodiment as a figure of speech. This is the way the term "embodiment" has been most commonly used by those practised in the linguistic turn. Some readerly caution is required here to sort out what authors intend because, for those of us who are seeking ways to find out about technologies, environments, and everyday practices materially and directly rather than by depending on metaphors and other figures of speech, embodiment employed as a figure of speech (i.e., epistemological embodiment) is a less capacious concept than the corporeal.[28]

Bodily encounters with technologies, environments, and everyday practices also occur profoundly through the senses of taste, touch, and smell, the senses Barbara Duden separates out as knowledge held "beneath the skin," and those Elaine Scarry, in her studies of pain, characterizes as "resisting representation."[29] These senses, with proprioception, kinaesethetics, and proxemics, are less welcome by scientists seeking evidence and by science studies scholars seeking conceptual analogies specifically because they resist being represented (and simplified) in symbols and models. Physicians in the early modern period, participants in the making of science in their time, shifted from using internal to using external signs of illness in their diagnoses and came to less often ask their patients "how are you" – meaning what can you yourself sense in changes inside your body – and to more often assert, "I can see how you are." This change was radically elaborated in the twentieth century. Diagnosis by physical examinations that used touch (through palpation), smell, and hearing (through stethoscopes) became less central than electronic apparatus that yielded graphic output that practitioners could see and compare.[30]

But these scientific and science studies simplifications, which attend mostly to the external senses and set aside bodily knowledge in favour of models and ungrounded theory, have come at some cost. For those of us seeking to know the full complexity of the relationships among humans, their environments, and technologies, these material manifestations through the body are indispensable. For us, they are not "experiential excess" but, rather, are key to what we need to know about human interactions with their tools and physical settings.

The processes of corporeal embodiment have histories. As human interactions with environments, technologies, and the everyday have changed, the senses have been tuned, over time, to bring different qualities to human bodies. The most commonly cited example of this is the presence of moveable type, which made people more dependent on their eyes and, as they became less reliant on their ears, less practised listeners.[31] Because the senses have been retuned by human experience with technology and the environment, treating the differences between them as separate, distinct, and persistent over time would be a mistake.[32] Thus, the leading practitioner in the history of the senses, Alain Corbin, after producing elegant monographs on odour and sound, turned to integrated studies of the sensing self in time and place.[33] He wrote ecologically about the full-bodied experiences that nineteenth-century people had with the seaside and embarked upon a multi-sensory project to follow Louis-François Pinagot, a nineteenth-century craftsman,[34] through the "natural and social landscapes" he inhabited, emphasizing not difference but synaesthesia, the qualitative commonalities and shared conduits of the sensing body. Recovering these different ways of recognizing and organizing knowledge of the world, the accumulation of specific actions in specific places,[35] can be particularly valuable to us as students of the relationships among people, technologies, and environments. To do this, we must pay attention to the "complex specificity of human bodies," in themselves researchable legacies of sensation, "not merely products of discourse or objects of institutionalized power."[36]

The next part of the "how" involves creating more robust material histories of technologies, environments, and everyday life, after we have become aware (1) that bodily interactions with the world include unconsciously embodied senses (such as proprioception, kinaesthesia, and proxemics) organized by the size and capabilities of human skeletons and musculature; (2) that the five senses (sight, hearing, touch, taste, and smell) differ in how readily they can be expressed in words; and (3) that, historically, as they are refined by technologies and activities, we can find a more encompassing way to characterize the complex flow of embodied knowing.

Katherine Hayles built upon the insights of the Chilean cognitive psychologist Humberto Maturana who, before her, had emphasized that "the processes involved in our activities" constitute our knowledge and who recommended attending to the particular "relational and operational spaces" of living systems as material matters of doing rather than separating sensations, perceptions, and cognitions.[37] Hayles characterizes the bodily registering of interaction with the world beyond the skin as a "flux,"[38] as commonly constrained by the physical form into which human bodies have evolved, by the angle and acuteness of human vision, the speed of human pace, and the places and practices experienced by those we study. Hayles' rendering is that we have awareness "before conscious thought forms."[39] This is a stark departure from the conceptual frame of recent studies of the social construction of gender, sex, and race. Rather than postulating, as social constructionists have done, that meaning precedes experience[40] and that humans know the world through the meanings they share symbolically in language, Hayles, Maturana, Francisco Varela, Paul Connerton (with Pierre Bourdieu, Lefebvre, and Maurice Merleau-Ponty)[41] suggest that humans "make sense"[42] of the world, of technologies, environments, and the everyday, directly through their sensing bodies. By storing the consistencies in this awareness, humans become habituated to their habitats, comfortable as practised users of their tools,[43] and share what they have learned directly by imitating one another. In this sense, humans are not first language bearers but, to use Varela's term, embodied minds.[44]

If I have been doing my job well, you now have some understanding of why we as readers and writers of environmental history, the history of technology, and the everyday, have much to gain by attending to how bodies learn and reason about their habitats and tools. Embodied perspectives allow us to tap into more of the knowledge humans have and to learn more about the reasoning they employ as they use technologies in the places where they dwell, work, and play. Their tacit knowledge – Douglas Harper calls this "working knowledge" in his fine study of a skilled Saab automobile mechanic in upstate New York[45] – is key to capturing the complexly intermingled texture of human experiences and technologies. Similarly, humans' sense of place, how they come to feel at home and sense danger, how they adapt so as to be competent when moving through the environment and being attuned to its changes, comes from the processes of corporeal embodiment. These are the ecological senses that William Cronon recognized were being marginalized by the "rhetorical razor" of the linguistic turn.

Science and technology proceed "hand in hand" – this, I hope, you'll now recognize as a use of embodiment as a figure of speech – but they are not the same. As readers and researchers in environmental history, the history of technology, and the everyday, we have much to learn from hydrology and metallurgy, from biochemistry and soil science. But many of the simplifications these disciplines employ to design empirically verifiable experiments and to organize their findings through theories derogate knowledge and reasoning not readily represented in symbols and signs. Thus, these practices set aside too much of the robust materiality humans learn from direct bodily interaction with environments, technologies, and the practices of everyday life. To reclaim a more complete understanding of these common and profound parts of daily human experience, we need to open interpretive and analytical space for the corporeally embodied knowledge that resists representation in language. My argument is that, if we do this, our environmental histories and our histories of technologies and daily life will be more rich, complex, nuanced, and useful.

The next part of the "why" concerns an absence that the reader may already have suspected. William Cronon's 1992 concerns about the "rhetorical razor" have prompted caution among environmental historians employing discourse analysis. Adam Rome, in one of his last writings as editor of *Environmental History*, urged more attention to research on the senses as ways of knowing the environment.[46] Two recent articles in that journal, one by Peter Coates and one by me,[47] are evidence of some opening up to work that attends to corporeal embodiment. But the scholars who have done the most cogent thinking about the processes of corporeal embodiment – to my mind, Hayles and Hansen – have not explored its historical manifestations. Hayles even castigated Hansen, in her foreword to his first book, for the "remarkable absence of particular technologies used either as examples or as occasions for analytical exploration."[48]

Writing in 1995, Isabel Hull, a former editor of the *American Historical Review*, argued that the knowledge and reasoning I've been calling "corporeal embodiment" were particularly amenable to exploration by "self-reflexive, scrupulous historicism,"[49] and that following the practices of the historian's craft we could reclaim these important but recently marginalized elements of human experience. The second part of the "why," then, is that historians are particularly well equipped, thanks to the closely textured integrating and contextualizing conventions of their craft, to open up this interpretive terrain for our own purposes, to model practices that colleagues in environmental and technology and broader social science studies might emulate. And so, the theory dispatched, let us turn to some examples from the recent literature that focus upon what corporeally embodied experiences with technologies and environments can reveal.

For the sense of safety, corporeal embodied knowledge is particularly important. True, people in the last century learned how to be safe at home, work, and play by reading manuals. Governments and firms developed occupational health and safety documentation for many workplace processes, and manufacturers provided instruction booklets to accompany the domestic appliances and home gardening equipment they sold.[50] These were textual ways for employees at work to augment their shop floor knowledge and for novice users at home to protect themselves from injury in their houses and yards. But this participation in what discourse scholars call "interpretive communities" was not sufficient protection against hazards.

Keeping safe, at root, is a matter of tuned bodily practice. Especially in risky environments, made dangerous by the presence of potentially toxic technologies,[51] well-honed, unconsciously held reflexes are key to human safety. In order to be effective, these responses to warning sensations must be embodied and automatic. Barbara Allen describes the embodied knowledge of place among families who had lived along the Mississippi since their great-grandparents had been emancipated from slavery. During their grandparents' time, these places had been altered by the advent of oil refineries, petrochemical plants, and aluminum smelters. These people understood the land as "constitutive of who they [were]," and they knew their survival depended upon practised physical proficiency in traversing the landscape in order to escape. These were their bodily legacies of persistence. When alarms in the night signalled a chemical leak, a gas release, or an imminent explosion, safety depended upon residents' intimate visceral knowledge of the night landscape sensed underfoot. By following the route, copying the gait and carriage of a leader, staying upright and moving forward, they practised a bodily tactic, using embodied knowledge to resist a bodily threat.[52]

Some embodied knowledges are particular adaptations to specific bodily work histories, the corollary being occupational demands on households. If learning an occupation is, in part, the development of reflexes that tame industrial hazards into definable threats, the rhythm of the workday and the work year require emotional and logistical accommodations on the part of families. The households of workers develop an emotional compass honed to flexibly, or at least stoically, accommodate dad's being away or mom's being on night shift. These habits, skills, senses of comfort or dis-ease make recompositions of the household variously supportable or unsustainable. They can be material influences of moments in histories of technological change, even as they elude the "human resource" evaluations of credentials.

Miriam Wright found that planners for the modern offshore fishery in Newfoundland sought out young men from communities where the household-based inshore salt fishery was in decline. They expected that the high levels of literacy, specifically the ability to learn from technical manuals and familiarity with the pace of factory work, governed by clock time, found in these communities would be key to trainees' successful transition to the trawlers of the new industrialized frozen fish sector. However, Wright discovered, if their book-learning was an advantage, the senses of self these men embodied as workers and fathers/husbands made them unsuitable candidates for the technologies and environments of the new offshore fishery. The workplace dangers they had learned to handle were the proximal threats of inshore waters, of near rocks and winds deflecting around a ragged coast. The hazards of the open seas of the offshore challenged their proxemics, more unconsciously than consciously held, and their learned sense of appropriate spatial scale in their work environment. Moving from the inshore to the offshore meant exchanging the comfortably near for the threateningly far, and it entailed a fundamental rupture in their embodied sense of space.

These sources of unease were amplified by the stresses upon the gender and household division of labour caused by men being away for weeks at a time. The workers who fit readily into the new industry were older

Banksmen, who often had little schooling but who were long accustomed to working offshore on the high seas and who came from communities long adapted to the absence of men for extended periods. These men recognized themselves in the relational and operational spaces of the off-shore fishery; their wives had been raised in traditions that managed the boundaries of psychological, physical, and intimate space so as to accommodate the departure and return of adult men. It was these customary, often unconsciously held embodied affordances, key to manly and womanly competence, that distinguished former inshore fishers from former Banksmen (and their families too) as differently adapted to the new industrial environment.[53]

Sensuous tuning, like the inshore fishers' assimilation of the winds and currents of the coast and the offshore fishers' adaptations to the routines of the trawlers and to the storms of the high seas, is referred to by Thomas Csordas as "modes of somatic attention."[54] Raymond Smilor finds,[55] similarly, that humans who moved from rural areas into towns were initially beset by sensuous challenges that required new modes of somatic attention. The country person who, in order to live off the land, had mastered a quite finely honed discernment of bird and animal life, and a refined awareness of subtle changes in the sky and the shifting winds, was overwhelmed by the sensuous barrage of the city. Rural migrants adapted to the city by developing sensory calluses – in effect, physical habits of inattention – in order to bear hubbub and function satisfactorily amid the surrounding roar. These new, and higher, bodily tolerances for the noise, stench, dirt, and shadow of congested neighbourhoods were gradually learned "shifts in perception," adaptations to losses that were made acceptable by the promise of material gain. In the technological and ideological regimes of the late nineteenth and early twentieth centuries, this sensory rehabituation passed into symbolically codified regulatory regimes that set down the tolerances of the "person of ordinary sensibilities" as the standard. Such municipal and state initiatives were recognitions that citizens had honed differing habituated sensitivities. If diverse populations were to work, live, and play together in closer quarters than they did in the countryside, such a median sensuous standard needed to be defined. Once these standards were established, landuse zoning was implemented to spatially cluster activities that posed similar sensory challenges. Additionally, acknowledging that the bodily assaults and pleasures of urban life were varied and that the money-making or pleasure-seeking of some might unreasonably discomfit the "ordinary sensibilities" of others, municipalities developed scales of fines based on rankings of how loud, stinky, or dirty activities were in comparison to what a reasonable citizen should be obliged to bear within urban boundaries.

In our own time of personal sound technologies, money can buffer these differing embodied tolerances. Those who want loud music can clamp speakers directly onto their ears, and those who want silence can choose to use noise-cancelling earphones against the rising din and the increasing deafness of their neighbours. But declining air quality and rises in global temperatures are different. These pose sensory challenges to the embodied routines and postindustrial quality of life expectations, and for them we have to turn to social, regulatory solutions rather than to individual remedies.

Technologies, too, form and frame the sensuous flux that flows through bodies. Virginia Scarff has written widely and with elegant gusto about how access to automobile technology changed human senses of the western American states from places to be in to places to pass through. She makes us feel the differential impact of access to transport on the lives of women and men, what registered of the places at speed, from the seat of the car, perhaps with the radio on. Then she parks and shifts focus and scale. If being in the vehicle gave female drivers access to more places, Scarff notes, being fashion conscious about footwear and underwear constrained how they moved and how they breathed, intimately reordering their perceptions of distance and slope, of the surface beneath their soles, of what they could bend to reach, of what a flirtatious impulse exposed and what modesty forbade.[56]

Isaac Asimov argued that if the automobile had not been invented, modern life would be far more different than it would have been had Einstein not articulated the theory of relativity.[57] Scarff's work gives texture and force to his contentions. We make sense of our habitat through the sensations our technologies and actions allow, at least as much as we do through conceptual frames, symbols, and signs. By contrast with Scarff's women drivers in the American west, consider what walkers learn about their environment. To amplify this contrast, imagine a walker who is not wearing earphones.

There are two fine books about the history of walking that I recommend, but my own research may also help make the point.[58] To make room for the St. Lawrence Seaway, eight colonial hamlets were relocated back from the St. Lawrence River. They were designed and resettled as three modern villages. Almost everyone with whom I spoke in the new village

of Iroquois claimed that the nineteenth-century settlement they left in
1958 was more interesting than the modern streetscape of crescents and
cul-de-sacs, the plaza, and open parkland planners designed for them
1.6 kilometres north, beyond the reach of the flood. People walked in the
old village; in the new village they drove. Can I know, then, whether
there was less texture and variety in the modern town than in the Loyalist
village or whether a change in the technologies through which they moved
about the former separated them from the sensuous contact that in the
latter would have given variation to their daily lives? I think this is a good
historian's question, especially for those of us interested in environments,
technologies, and the everyday. It is a bigger query than our customary
historiographical questions, for it goes beyond analysis of the content of
the discursive form, and the message in the medium of the technologies
we use,[59] to take the full-bodied sensation of being in and moving through
place, mirroring the scale and qualities of the location, as direct experiential evidence of the content of cognition and structure of reasoning.

Consider how a road system, surveyed before the railways upon a square
grid, might function as another technology constraining bodily encounters with the environment. Travellers on these roads meet the place through
its property relations. Their route follows not the topography but the road
allowances, not the most breathtaking or the safest route, but the route
least likely to prompt charges of trespass. Whereas the crow flies, the
stream flows, and the wind blows in movements responsive to the lay of
the land, travellers respecting private property cannot follow either the
shortest or the least physically demanding route. They move not in bodily reciprocity with the landscape but along a path that minimizes their
intrusion upon the rights of property owners.[60] Perhaps they learn more
of the township from their journey along the surveyed road allowances
than an air traveller passing through an O'Hare terminal learns of Chicago; perhaps the person transiting through O'Hare is merely more
conscious of the technologies that bring the world from beyond the skin
into human bodily understanding of near and far, fast and slow, hostile
and homey.

If movement from the nearby countryside leads inhabitants resettling
in town to develop sensory carapaces, literally hardened sensory shells to
armour themselves against the assaults of their new habitat, people leaving
home to cross continents and oceans find their knowing and being in place
assaulted in at least as many dimensions. Studies of migration and resettlement, at their best, recapture the "sense of physical vulnerability"

of newcomers struggling with a somewhere they have had not yet viscerally embodied, where familiar habits do not serve and accumulated tacit knowledge cannot connect.

Sometimes, as Thomas Dunlap finds in studies of British settlers in Australia at about the same time, populations "could not even think what they wanted," so disparate was the place where their bodies were from the place their bodies knew. They were deracinated, their roots taken from the soil upon which they had learned to thrive. And thus they embarked upon a project, using imported plants, animals, technologies, and building forms from the continent they had left, to remake the threatening new territory of Australia into a simulacrum of the English countryside. Like the artifices made virtually by aficionados of the contemporary computer game SimCity, the result was a copy for which there never had been an original, a fantasy made by longing for a different space rather than by scrupulous close encounters of a daily kind. Some among the first generation were made abject by the loss of visceral correspondence between self and place. Yet humans can, and do, habituate to new "natures." They develop new reflexes and embodied understandings of nature to mirror their changed environments. To the children of the second generation, Dunlap notes, "Shelley's nightingale and English hedges were as alien as the woollen school uniforms they wore in Sydney's heat," for the familiar these daughters and sons had embodied as natural was not their British-born parents' first familiar but, instead, the ordinary of their part of the southern hemisphere.[61]

Carolyn Merchant notes a similar difference in what Native Americans and Euro-Americans recognized in North America. She argued that Native Americans knew their environment as a tame and bountiful dwelling place. They experienced the earth as "an agent of regeneration," where women planted the corn, beans, and squash in forest gardens, and the produce of these activities sustained life. They reasoned about nature as an ascensionist and progressive force in their lives. Euro-Americans, by contrast, encountered the New World through the seventeenth-century concepts of Christian religion, modern science, and capitalism, which led them to consider the environment as "postlapsarian," a desert made when Adam took Eve's poor advice in the Garden of Eden. People's job, thereafter, was to seize control of nature and recover what had been lost. The manner in which Euro-American traditions epistemologically embodied nature made them think of their harvests as the fruit of their own labour and proof that God favoured them rather than as the product of a partnership between human and non-human nature in which humans by corporeal embodiment had discerned nature's autonomous ways.[62]

Partners function best when they openly attend to one another. Equally, in the wilderness and in the garden, William Cronon urges us as humans "to recognise and honor nonhuman nature" as a world with its own independent, non-human reasons for being, "whose otherness compels our attention."[63] This means recognizing not only nature as represented in models of empiricist science and theological doctrine but also nature as the autonomous force that sensing humans corporeally embody as "a feeling for the organism."[64]

Scholarship on the social construction of gender relies heavily on embodiment as a figure of speech, and it is framed to understand the meaning of gender formed in power relations and institutional structures. This is how Judith Butler analyzes the aetiology of "gender trouble" in her book of the same name, and, in *Bodies That Matter*, it is the basis for her insights into how contemporary gendered assumptions constrain the research questions modern geneticists ask about sex.[65] Joan Scott traces the discursive history of "gender as a category of historical analysis" from similar epistemological premises.[66] But gender is also created by active experience, by the differing corporeal embodiment that flowed from the different labour women and men have performed historically, the differing places in which they lived, worked, and played, and what their daily lives taught them about the nature of their environments as active forces or terrains to be mastered.

Consider how, in *Silent Spring*, Rachel Carson's depiction of nature as active and responsive altered conceptions of biology, culture, and environment.[67] Through her texts, Carson provided a way for urban moderns to share a sense that planters and gleaners had long embodied through their techniques[68] – that selves existed in reciprocity with nature, that humans were not nature's masters, that nature could bite back.

When humans develop gendered sensibilities, they do so through "species specific, culturally formed and historically positioned" experiences.[69] Vera Norwood notes that, among both birds and humans, gender roles differ. Men are more likely to chase, capture, and signal, while women are more likely to watch, nest, and mend. These gender roles are critical to understanding how humans construct nature. "Nature is used to define human nature" and, thus, to endow as "natural" the questionable privilege "granted to some embodied positions over others."[70]

The same gendered relationships between doing, understanding, and being emerge in some histories of toxic technologies and the environment.

In a wide-ranging 1992 review of research on environmental concern, for example, Paul Mohai noted a "gender gap" between the activism of women and men.[71] When Phil Brown and Faith Ferguson followed Mohai's work with a study of women's work and women's relationships in toxic waste activism, they found patterns of doing and knowing best explained as outcomes of differently corporally embodied experience and knowledge.[72] Brown and Ferguson begin by confirming that "grassroots activists who organize around toxic waste issues most often have been women, led by women." They then note that women's situations in family, work, and place positioned them daily and concretely to perceive environmental effects, particularly on the health of family members whose physical care in time of illness was often a female responsibility. Through intimate contact, feeding, bathing, and toileting the distressed; through the internal revelations that arose through these direct bodily contacts – touching, smelling, and hearing – they tracked the presence, process, and growing intensity of the environmental body burden. The knowledge they gained was corporeally embodied, the experiential product of monitoring internal changes, and it was not manifestly comparable or readily shared. The very relational capacities that made them open to the flux of suffering disqualified them as autonomous sources of knowledge and discounted what they learned as being empathetic rather than objective.[73] The challenge for these lay toxic waste activists, both women and men, was to "blunt the rhetorical razor" and insist that what they knew was not "experiential excess" but crucial to understanding how the toxic outputs of some contemporary technologies was altering the environments in which they lived, worked, and played.

If embodied histories are key to understanding how humans have kept themselves safe, how they have honed skilful practices in order to interact with the world through technologies, how they have recognized the environments they entered and subsequently reorganized (in the process remaking themselves as sensing beings), bodies are also places and repositories of histories of practice in place. Three recent books in environmental history elaborate upon these relationships. Gregg Mitman concludes his study of allergies by affirming, "Place matters: in the making, in the experience, and in the knowledge of illness and health."[74] Thus, he challenges the commonplace notion that health is centrally about the absence of pathogens and the accessibility of miracle cures. He asks us to also consider health as an ecological relation, to make conceptual space for the aspects of well-being that elude the linearity of medical models, that emerge from the diffuse meeting of the body with what Duden calls the world "beyond the skin," what Hayles refuses to partition by her choice of the term "flux." Linda Nash calls these relationships between physical human bodies and the larger world "inescapable ecologies."[75] She shows how European Americans felt so out of place in late nineteenth-century California, so unable to resolve the discrepancies between the natural order they had embodied and the "nature of things" where they now were that the pioneer imperative "to conquer nature" was incomprehensible. This was territory to be discerned and "completed," which could then be only imitated rather than remade, but which twentieth-century industrial agriculture remade so thoroughly into hydrids so complex and unstable that they eluded the "reductionist methods" of scientific expertise. Her histories of the Imperial Valley reveal instances in which laypeople, less forcefully disciplined to discount or "forget" awarenesses learned and made useable by daily dwelling, continued to have access to the ecological relationships in their industrialized bodies and the industrial environments beyond their skin. Similarly, most of the people about whom Conevery Valencius writes,[76] nineteenth-century inhabitants of the Arkansas and Missouri borderlands of the American South, recognized that "the external world and the human body were not as separate as they are now" and acknowledged the "constant interplay between human agency and human lack of control, rather more frankly than we moderns." Valencius calls corporeal embodiment "physical citizenship," changes through which the bodies of migrants "would be 'acclimated' to new climate and topography" in a struggle "both crucial and perilous," which simultaneously accommodated external geographies and forged anew the "complicated interior geography of sensation movement and flow" that "determined well-being." She describes how the migrants to the borderlands retuned their sensing bodies to the sounds, tastes, textures, and spaces, the "lived realities of the terrain," learning to make sense of "the nature of soil, water and situation," remaking the embodied minds that connected them, individually and collectively, to the world beyond their skin.

These processes of corporeal embodiment, forged individually in practice and collectively by imitation and shared practice with technologies and environments in the routines of everyday life, are the focus of the six case studies that follow. Often corporeal embodiment and embodiment used as a figure of speech exist in a tandem circuitry. Henri Lefebvre described this relationship as a masque of "reciprocal implication and explication," in which the material and the social body converge and then separate, presenting alternately as real and illusory, palpable and absent.

Sensuous knowledge, ecological awareness, can be told. It sometimes withstands the operations of the rhetorical razor and defies derogation as experiential excess. But in our times, embodied histories are often repressed or forgotten. The task here is, through the interrogation of several vibrant and vital instances, to reclaim them.

http://
megaprojects.uwo.ca/Gagetown

This site contains the new media work associated with the creation of the Gagetown NATO base upon the former farmlands and woodlots in the Nerepis Valley in New Brunswick.

The aerial and satellite imagery on the website make starkly apparent the landscape changes in the Nerepis Valley since the opening of the Gagetown base. Using satellite imagery from Google Earth, topographical maps, and aerial photos from the National Air Photo Library in Ottawa, this site shows vehicle mazes appearing and tank tracks crossing the land of the Scotts and their neighbours, succeeding earlier roads and paths. Equally striking are the tree lines and land distinctions *persisting* as physical registers of prior landuse practices.

There are also photographs and audio material that portray the local residents and their former ways of life: Lydia and Raymond Scott, their farmlands, Jerseys, and the hooked mats they made together. Multimedia material gathered from the anniversary tour of the area held in 2003 is available, including images of the barren lands visible from the bus with white placards standing in for places since removed as hazards. Lastly, there are audio and videos of music and dances collected by Greg Marquis and Donn Downes of the University of New Brunswick, Saint John. These are enduring signs of continuity amid the disruption of the ways of life of the Scotts and their contemporaries in the Nerepis Valley.

—JV

PLACE AND CITIZENSHIP

Woodlands, Meadows, and a Military Training Ground: The NATO Base at Gagetown

L and for the Gagetown NATO training base, the earliest of these six post-Second World War case studies, was assembled in 1952 from farms and woodlots in the region of south-central New Brunswick back from the port of Saint John. There the lower Saint John River Valley is wide and welcoming, its fertile and well-watered meadows, as a local daughter recalled, "sloping up to high rocky ridges where once grew magnificent stands of pine."[1] For at least three centuries, the region had been a place of succeeding human population, refeatured by the currents of North Atlantic geopolitics, marked by the legacies of exile and immigration. In his reflections on his own history of marginality and displacement, Walter Benjamin, who fled Germany for France and during the Second World War died as he struggled to flee France for Spain, captures something of the forgetting and refusal to forget such a place as Gagetown begets: "To articulate what is past does not mean to recognise 'how it really was.' It means to take control of a memory as it flashes by in a moment of danger."[2] At the height of the Cold War three thousand people, 834 families, were moved off pastures, meadows, fields, and stands of timber long and carefully tended (and never intended for sale) to make way for tank mazes and firing ranges.

Let us begin by acknowledging the very allusiveness of personal histories in place: that being forced from home, from serviceable habits in a habitat viscerally known, challenges a present haunted by past vestiges.

In 1604, the French joined the First Nations inhabitants of the region. Through the eighteenth century, these Acadian settlers around Grimross (contemporary Gagetown) found themselves, through the fortunes and misfortunes of war, by turns the subjects of the British and the French Crowns. When, under the Treaty of Utrecht, 1713, the French ceded Acadia to the English, these formal European imperial alternations ended. Then, in circumstances that foreshadowed and shadowed later events, for a long generation the Acadian inhabitants of lower Saint John remained at the pleasure of the English Crown, shaping and being reshaped by the terrain, until in 1758 and 1759 they were compelled to flee before British troops implementing a new North Atlantic alignment.[3]

The Acadians were replaced by New Englanders, who were called planters and who were recruited to take up the farms and woodlots from which the French-speaking Roman Catholics had been expelled. In 1784, New Brunswick was created as a separate colony for Loyalist refugees from the American Revolution. Loyalists were granted the best land along the Saint John River. The Acadians, returning, settled in the distant north of the colony. Through the 1820s and 1830s, impoverished artisans and agriculturalists from the British Isles, principally Protestants, took up the first and second tiers of land back from Gagetown. In the 1840s, they were followed by Roman Catholics fleeing the potato famine in Ireland, who moved into the higher lands farther back from the front. Base Gagetown, a creature of the Cold War, was set down in a place made by four hundred years of jostling between French and English, Catholic and Protestant – an imperial genealogy of settlement, displacement, and resettlement, allusively present and absent, which gave the sense of being at home a sharp defensive edge.

The Canadian Army, trained for the Second World War overseas and reduced in 1950 to a small peacetime force, trebled with the decision to enter the Korean War and the escalation of the Cold War in Europe. NATO obligations required not only a larger standing army but also an all-weather training area to accommodate a division of fifteen thousand men, the safe use of new longer-range artillery, multiple mortar and rifle ranges, grounds to train drivers of tracked and wheeled vehicles, and space for target practice for tanks. In August 1952, the Canadian government announced that an area fifty-two kilometres square by seventy-seven kilometres square northwest of Saint John had been selected for this purpose.[4]

The area back from Gagetown was chosen for this military training ground because it was on the Atlantic seaboard, near railheads and the major deep-water port of Saint John. Sparsely settled, yet with a sufficient

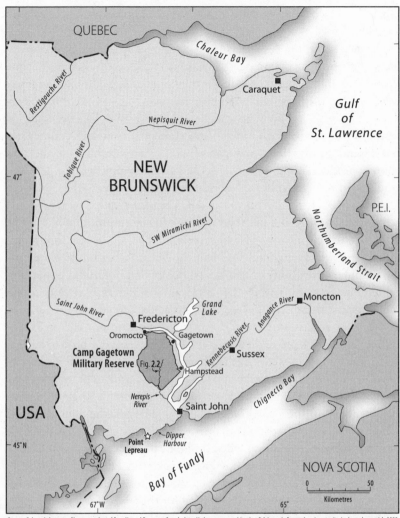

FIGURE 2.1   Southeast New Brunswick, showing Base Gagetown, drainage, and large urban centres | Cartography Section, University of Western Ontario, Department of Geography

legacy of habitation to simulate urban street fighting, and conducive to field training in all seasons, these lands included a "variety of notable terrain" admirably suited to the army's purposes. Major-General J.M. Rockingham, named the first commanding officer of the camp, was pleased with the site: "It's got mountains like Korea, lowlands like the north German plains, mud like Italy. You could train men here to fight anywhere."[5]

Defence spending was crucial to the regional economic development plans that the Canadian federal government promised after the Second World War. By the end of the 1950s, a quarter of all Canadian armed forces were based in the Maritimes, and military spending was nearly $200 million annually, a vital boost to a region whose fortunes had been lagging behind those of Central Canada throughout the twentieth century.[6] In the summer of 1952, local newspapers were filled with tallies of the agricultural produce, building materials, and civilian employees the new base would require. Though the question then asked by ninety-three-year-old Elizabeth McCarthy – "What price do they pay for contentment?" – would echo down through the next half century, in the community halls of an area with a long history of Loyalism and military service, audiences quietly absorbed Department of National Defence allusions to Korea and the imminent possibility of another world conflict. John Hansson, interrupted that September while harvesting a waist-high field of grain on his modern farm at Armstrong's Corners, near a home recently equipped with all the latest city conveniences, summarized the Hobson's choice: "If the country needs this place, they need it. We don't want our boys to have to go to war again, do we?"[7] In the economic and geopolitical context of the time, the military necessity and the economic benefits of the base were unquestionable. Leaders from the affected region – the Orange Lodges, the women's institutes and the agricultural societies, the physicians and local members of the federal Parliament and provincial legislature – could only ask why, in a province where millions of acres remained unsettled wilderness and vast sums were being spent to rehabilitate dykelands for farming, so many excellent long-established farms and so much future agricultural revenue should be sacrificed for the base.[8]

Persistently present in this group of community leaders were Lydia Clarke Scott and E. Raymond Scott, a couple whose families had farmed and managed timber lots for many generations near Summer Hill in the Nerepis Valley along the Lawfield Road, which ran up the centre of the proposed base from Armstrong's Corners to Gagetown. In July 1952, Raymond, with other leaders of District Farmers Association No. 6, drafted a petition to Canadian prime minister Louis St. Laurent, describing "our farms, our means of livelihood and a well established form of living" as being beyond adequate compensation.[9] Their claims were grounded physically in an embodied form of living and polemically in analogies between the eighteenth-century expulsion of the Acadians and the mid-twentieth-century displacement of similar numbers for the military base. That year in a provincial election, the Conservative Party, which commanded Orange support, defeated the Liberals, with whom francophone New Brunswickers were traditionally allied.[10] Three years later, denied access to the public broadcaster and the station owned by the *Telegraph-Journal* in Saint John, the Scotts and their allies among the District Farmers Association made a last stand in a series of evening radio addresses on upstart CJBC Saint John, describing their home places, the woods, and the lands, the community institutions, and the women's world from which they could not depart without leaving parts of themselves behind.[11]

What were the practices of everyday life, the embodied "well-established form of living" that these farm families made of themselves and their place on holdings back from Gagetown? Theirs were not wilderness lands in 1952. Already a century before, British visitors praised the fine farms along the Lawfield Road and the surpluses from which they provisioned townsfolk of the port of Saint John.[12] The men of the Nerepis Valley were farmer-lumberers, inhabiting what is called in the Acadian and Quebec literature an *économie agro-forestière*.[13] New Brunswick had been at the centre of the nineteenth-century square timber export trade and, into the 1940s, was the Canadian province most dependent on the forest industry.[14] When the decisions about Base Gagetown were being made, the leading Canadian forest historian, A.R.M. Lower, characterized New Brunswick as the place where the relationship between forestry and agriculture was most fraught, where timber stands were consumed by fires in the slash that lumberers prodigally left on the forest floor, and agricultural development was impeded by the competing demands for labour in the woods.[15] Since then, scholars have described this picture as "overdrawn," as a depiction coloured by the homilies of the nineteenth-century agricultural press; rather, they conclude that the mixed hardwood forest prevailing in the Lower Saint John Valley was being culled and that small woodlot owners in the Maritimes, aiming "to produce a fairly open multi-aged woodlot," had historically practised sustainable forestry.[16]

Some species in the woods of the Nerepis Valley, technically classified as Acadian forests, grew as rapidly as any in the province. Individual trees could live three hundred to four hundred years. Among the stands of tolerant hardwoods – sugar maple, beech, and white birch – grew hemlock, pine, and spruce, the conifers more common along the Lawfield Road on

steeper slopes of Coote Hill Ridge, Headline Ridge, Inchby Ridge, and
Jerusalem Ridge as well as in wetter sites below. These resources were
attentively husbanded, for they provided cordwood for fuel, hardwood
and pulpwood for sale, maple sap and fiddleheads in spring, salmon in
summer, and venison for the freezer in fall.[17] In what would become the
Base Gagetown training area, agriculture and lumbering were comple-
mentary. A contemporary study found that farm families in the area received six-tenths of their cash income from agriculture and one-third from
the proceeds of their woodlots. Five out of six had no indebtedness in
1952.[18] Even as dairying intensified after 1930 and this part of Queens
County became key to the milk-shed of Saint John, farmers expanded
their woodlots.[19] In the interwar years, a growing demand for pulpwood
increased incomes on mixed farms in south central New Brunswick by
$100 per year.[20]

Though lumbering was practised as a winter activity, recollections
emphasize the associations between the woods and warmth, both bodily
and social.[21] In late fall or early winter, male neighbours went with their
sons into the stands on the highlands at the back of their holdings to cut
firewood, with their horses yarding the felled trees into piles called, perhaps
for their shape, brows. Once snow cover made hauling easier, the group
returned to each farm in turn to draw the logs by horsepower into two
brows side by side near the house. Come spring, they returned with a
sawing machine and a gas engine to cut the logs into eighteen-inch (forty-
six-centimetre) stove lengths. On these days, the women of the household
provided meals. One man recalled roast meat for dinner at noon and, for
supper, a scallop, with Washington pie or rice pudding for dessert. Thereafter, the families split their stove wood, stacked it outside to dry through
the summer, and come fall moved it back into the woodshed. Each household would require fourteen to sixteen cords for cooking and heating each
year. "Getting the wood in" consumed family labour for several weeks
annually, and the process has remained a necessity in the region, a seasonal ritual and an affirmation of providence.[22] In a time when other North
Americans were displaying their comfortable postwar circumstances by
posing for photographs in front of their automobiles, the families along
the Lawfield Road, though they owned cars, assembled for the camera
atop their woodpiles.[23]

By contrast, the lumber camps were a homosocial world where – the
exceptional female cook aside – men and boys, with their horses, spent
uninterrupted weeks and months in one another's company. The lumber

camps were a kilometre and a half or more back into the woods, primitive tar-paper-covered structures with a cookstove and eating area at one end and, marked off by a cloth partition, bunks made with boards and straw for "sleeping and telling yarns" around a Quebec heater at the other.[24] Deep in the woods there was no wind. Although later in the 1950s chain-saws and diesel-skidders entered the forests of New Brunswick, in the Nerepis Valley, until 1952, men and boys worked with axes and cross-cut saws, the quiet of the day broken only by the falling of trees and the move-ment of the teams. In 1911, there were 1,009 horses enumerated in the parishes that would become the base, and there were still 608 in 1951.[25] The food was plain – oatmeal porridge, beef, pork, root vegetables, prunes, beans, and tea without milk (never coffee). Although the men missed fancy cakes and pies, which were the pride of valley farm kitchens, they fondly remembered from their boyhoods the intense sugary delights of camp molasses and ginger cakes, and doughnuts dressed liberally with maple syrup. Besides the "telling of yarns," the evenings were filled with fiddle music (fiddles were brought to the camp and some were made in the shanties) and dancing. Lydia Scott's youngest brother, Willard, remembers that as a boy he "always had to dance as a lady."[26]

These accounts, mostly drawn from the recollections compiled for the Base Gagetown reunions of 1983 and 2003, respectively, are deeply sensu-ous – the narrators recalling the strenuous silence of the days, the company of the neighbours upon whom they depended and of the boys and horses they had trained, the smoky familiarity of the lore and the music, and, come spring, the good money. For the farmer-lumberers of the Nerepis Valley, the woods were diversified investments to be passed down the generations, dependable stores of annual income and savings against ad-versity, places of pleasure for their mingling of sensual satisfactions and prudent stewardship. For many among those who lost their holdings to the army training area, their exile from the woods was more brutal than their displacement from the fields.

This was particularly true for the Clarke men and, by his marriage to Lydia, for E. Raymond Scott. Lydia's father, Byron, was a millwright with a stationary engineer's licence. Early in the twentieth century, anticipating the responsibilities of fatherhood, he left his waged job at a Gagetown mill to build and equip a mill of his own on land he had inherited from his father back from the river at Hibernia. According to family lore, he sawed his first log there on 16 June 1909, the day his eldest daughter Lydia was born. In time, the Clarkes' woods business became a diverse vertically integrated operation that included a permanent lumber camp nearby. With his sons and a crew of a dozen or more, Byron Clarke sawed logs, planed lumber for dimensioned stock, and turned wagon spokes and handles for tools. Upstairs they milled grain.[27] By the time of the expropriation, Clarke owned 787 acres (318.4 hectares) of timberland in the base site. To each of his children he had given a hundred-acre (forty-and-a-half-hectare) timber lot.[28]

Lydia spoke of the wooded 323 acres (130.7 hectares) she and Raymond shared as a legacy: "We were really working on it and we realized it was coming ahead so wonderful. You could go into the lumber woods in winter and get enough to buy yourself a car or a piece of machinery or whatever you needed. A chesterfield, or whatever." By Raymond Scott's estimate, in 1955 there were 350,000 acres of rich woodland about to be expropriated for the base, stands averaging about 50,000 board feet (sixty-six cubic metres) of lumber per hundred acres (forty and a half hectares).[29] "Unlike other districts," a national reporter from the *Star Weekly* wrote in November 1952, "these woodlots had not been stripped, but carefully managed to yield steady yearly harvests." Frank Millar, the defence department supervisor charged to effect their expropriation, insightfully observed: "In other parts of Canada land is something to be measured in dollars and profit. There is a difference with these people ... they have a close attachment to the land they've lived on."[30] The woods were reshaped by their skilful tending, and from these practices and in their product – the stands of hardwood, pulpwood, and fuel – more than any other part of their holdings, lodged the sense of present competence and future financial security for the Clarkes and their neighbours.

From early days in the expropriation process, the future of the woods was in doubt. The Liberal premier John McNair, about to be unseated in the September 1952 election, assured voters that the owners of "wild lands" would have opportunity, over a reasonable period of time, to remove the merchantable timber from their holdings should they wish to do so."[31] "Wild lands," "over a reasonable period of time," "should they wish to do so" – the mind boggled. This assurance betrayed no understanding of the qualities of the timber stands, of the long time horizons, of the trusteeship over the resource the people of the valley had vested in the woods, or of their significance as sites of meeting and mutual refashioning between timber and flesh. The electors of the Lower Saint John voted Conservative, as was their custom, but for naught.

Later, press reports circulated that Ottawa had allocated $10 compensation per forested acre (0.4 hectares). Prices were high for "farm wood" from 1950 to 1952; the provincial average returns from the sale of forest products rose from $48 per farm in 1941 to $245 in 1951.[32] In the summer of 1952, when estimators hired by the Department of Defence were cruising the rich mixed stands of the Nerepis Valley, the annual returns from woodlots per farm would have amply exceeded the New Brunswick average, and commercial farms like that of the Scotts were probably closer to $900.[33] "Do the officials at Ottawa think this is a barren land?" Raymond Scott demanded over radio CFBC from Saint John. Lydia's brother Willard, a modest man and knowledgable woodsman, was confident that the army had received more from the sale of the remaining standing timber from his woods than they had paid him for the land.[34] In their calculation of compensation due, the Department of Defence placed little value on the forests on the lands that became the base. Like the people, the woods were in the way. In time, the woods did become useful, as testing grounds for chemical defoliants (at least in 1965 and 1966). Some of the area's displaced farmer-lumberers were hired on for wages to assist in these tests, a decision that many of them came to later regret when the health effects of working unprotected with Agent Orange and other members of the dioxin family became clear.[35]

By 1952, most families along the Lawfield Road had been on their farms for 125 years; the Loyalists and skeddadlers from the Revolutionary War, like the Clarkes, had been there for 175 years; and the planters had been there for nearly two centuries. They knew and had remade the land through generations of discerning observation, through learning accumulated as practice, and through experimentation with the innovations distant authorities had recommended as agricultural improvements. From 1910, members of the local Agricultural Society collectively purchased their grain and seed from a respected Upper Canadian propagator. In the Nerepis Valley, the agricultural boys and girls clubs, which introduced young people to the breeding, feeding, and management of livestock, antedated the 4-H Club. There were agricultural fairs in New Jerusalem from the 1930s and, from the 1940s, an illustration station supervised by officials from the federal Department of Agriculture. In 1949, the president of the New Brunswick Dairymen's Association came from Summer Hill, the community in which the Scotts farmed. Four men from New Jerusalem received degrees in the spring of 1952 from the leading Canadian agricultural school, MacDonald College, affiliated with McGill University in Quebec.[36] These were progressive farmers, schooled in the theory and practice of modern agriculture and skilled by long practice in the woods and fields they nurtured.

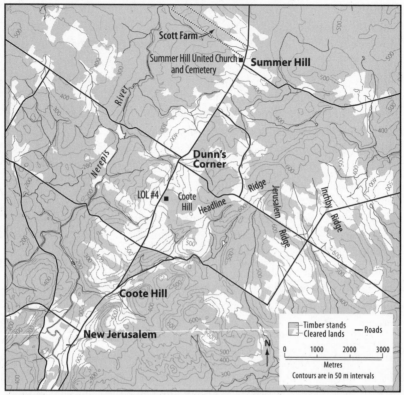

Sources: National Topographic Database. NTS Sheet #21G09 [computer file]. Sherbrooke, Quebec: Natural Resources Canada - Center for Topographic Information, 2009 and Hampstead, New Brunswick, Map 976A(21G9), 1/63,360, 1946.

FIGURE 2.2   Lawfield Road, showing Summer Hill, Dunn's Corner, LOL, and Scott Farm | Cartography Section, University of Western Ontario, Department of Geography

In the eighteenth century, Lydia's family settled on Long Island in Hampstead Parish, on meadowlands enriched by the spring floods of the Saint John River. After one flood too many, they moved inland, and by Lydia's grandfather's time had reached the second line of grants at Hibernia. The Scotts entered the area in 1818.[37] We know about agricultural practices along the Lawfield Road directly from the agricultural census and the accounts of Franklin Johnson, one of the 1952 graduates from MacDonald College, and, by inference, from Brian Donahue's richly detailed account of mixed husbandry on similar meadow and uplands in Concord, Massachusetts.[38]

Just as the richest agricultural land in the United States is not in New England, so the richest agricultural land in Canada is not in New

TABLE 2.1

Land use in the parishes of Gagetown, Hampstead, and Petersville, 1931-51

		1931	1941	1951
Acres	oats	1,847	2,104	1,404
	cultivated hay	9,326	13,601	8,243
	pasture	9,296	4,913	2,349
	natural pasture	3,469	5,746	–
	fallow	44	21	61
	roots	680	703	277
Number	cows in milk or calf	2,512	3,352	2,335
	milch cows/owner	5.1	7.05	7.6
	human population	3,362	3,313	3,290
Acres	cultivated hay/owner	19.3	28.6	26.7
	cultivated hay/milch cow	3.71	4.05	3.5
	cultivated hay and pasture/ owner	19.2	10.3	7.6
	cultivated hay and pasture/cow	7.4	5.52	4.5
	occupied	106,184	93,464	79,574

Source: Census of Canada, Agriculture, 1931, tables 37, 38; 1941, tables 51-54; 1951 tables 30, 31; and Census of Canada, 1951, 16-1, 17-1, 21-1.

Brunswick.[39] The soils of the farms lost to the Gagetown army training base were, by Franklin Johnson's informed account, "shallow, rocky and poorly drained with some outcropping of ledges in certain fields." In 1952, some had been under cultivation for almost two hundred years. In the twentieth century, the land under crop was mostly devoted to oats and cultivated hay. Farms also had many acres of fenced pasture and grazing land in prairie and natural pasture. These cropping patterns tell us a good deal about the strategies for prosperous persistence along the Lawfield Road.

The farm owners, 484 of them in 1931, 475 in 1941, and 307 in 1951, intimately knew the genealogy of the lands ceded to the Gagetown army training base. In the eighteenth century, when their meadows produced only scant forage of poor quality, farmers on the first and second tiers bought and hauled hay from the intervals and islands along the Saint John River. By the nineteenth century, specialization based on informal local observation and reference to the developing science of agronomy made this onerous practice unnecessary. The stony uplands of glacial till once logged off, though difficult to plough, remained verdant through the summer because they retained water. These were the acres of natural pasture for cattle that were recorded by twentieth-century census takers. The 1940s and 1950s aerial photographs of the land along the Lawfield Road show, in addition to these divisions, a well-developed infield-outfield system for crop rotation.[40] This system depended upon inhabitants' knowledge of their habitat. Their honed understandings, internalized as second nature, made their own practices a sustainable part of the local forest, field, and pastoral ecologies. In the nineteenth century, teamsters who hauled logs to mills and cordwood to Saint John brought back wood ash and, later, ground limestone – alkaline elements to ameliorate the acidity of the fields. Rotations among oats, grasses, legumes, and succulent roots for feed made the "shallow, rocky and poorly drained" fields described by Franklin Johnson produce abundantly.[41] Through the twentieth century, proprietors turned more acres from pasture to hay cultivation, keeping their growing herds of quality-grade Jersey breed cattle in barns and fenced yards nearby, where they collected the accumulated manure and reapplied it to the fields. Farmers fed pigs skimmed milk from their herds and sold the cream for butter to the local creamery.[42] In the 1940s, observing that commercial fertilizers increased yields at the local illustration station, many along the Lawfield Road supplemented manuring with this additional source of nitrogen. Through these several means, as royal commissioners found two decades later, local farmers maintained larger herds on their lands than did dairy farmers in the province as a whole.[43]

Except for the flooded meadows along the Nerepis River, enriched each spring by silt, the continuing fertility of back-country farms was not natural. Imbued with the intention of humans, the soil was a product of their planful attention, an extension of them. The alternative to this second nature, fields left untended returning to the "scant forage of poor quality" reported in the first years of settlement, was not forgotten by local history but, rather, was palpable and present as a warning. The Scotts' home place, bought by Raymond's father, Harry, "to go with his own" adjacent farm and left untended for a generation, was in this state when Lydia and Raymond assumed it in the 1930s. The hay crops that grew there were so light "you could look down the field and see through the stalks." Under Raymond's management, according to his brother-in-law Willard, the grasses became so thick "you could hardly cut [them] with a mower," and on their forty cultivated acres (16.9 hectares) that formerly "wouldn't keep one cow," he kept eighteen.[44]

The Scotts and their neighbours around Summer Hill were sophisticated dairy farmers who specialized in purebred Jersey herds. Lydia ran the dairy operation at the Scott farm and, by frequently making dinner for the local agricultural representative, Artie Hunter, she kept current about any science that might inform her practice. She turned part of their former infield into pasture so that the milch cows were near the barn, "right up front, this was their new plan," to save the half hour walk each way to the natural pasture on the verge of the woods.[45] Their local breeders' organization, the West Royal Jersey Club, which had its inception in 1942, had by 1951 exhibited more jersey cattle at its annual show than were shown at any similar event in New Brunswick. The group was provincially active in husbandry initiatives to improve the breed, and, through donations of heifers, contributed to the work of the Canadian Jersey Cattle Club.[46]

The West Royal Jersey Show held at Gagetown in August 1952 was the last. The fate of the purebred herds and their owners figured prominently in the November meetings that year between the District Farmers Association and the federal minister of defence. While buoyant contemporary markets for farm wood made it particularly disheartening to leave well-tended timber stands behind to decline, a contemporary ban on exports to the United States meant that 1952 prices for purebred cattle, which had to be moved or sold, were exceptionally low. Owners, who after expropriation were not going to farm again (among them the Scotts), faced a buyers' market when disposing of their cattle. The federal government promised additional compensation,[47] but money was not the whole of the matter. The breaking up of purebred herds left many in the valley both at a loss and lost. The heredity "built up" bodily in the herds manifested aspirations beyond profit-taking. The improvements made in their animals through breeding practices were accomplishments, evidence of skills in which the herders took pride.

From the loss of the herds followed the demise of local institutions that had sustained and been sustained by dairying in the valley, foremost among these the Hampstead Co-operative Creamery, founded in July 1947 after a meeting at Raymond and Lydia Scott's home, chaired by Lydia's father, Byron. The quality of the local herds contributed to the quality of the butter, and the Hampstead product took prizes in 1951 at the major Canadian agricultural exhibitions, the Royal Winter Fair, the Canadian National Exhibition, and the Ottawa Fair. The enterprise was expertly managed by Eugene Morris, a man celebrated long after the demise of the co-op creamery for his skill as a fiddler. The co-op also produced milk powder and sold feed, seed and fertilizer, home staples, farm supplies, and machinery. The Scotts bought groceries for themselves and feed for their animals at the co-op store and, after rural electrification in 1948, their home appliances as well. But because 75 percent of the butter fat the creamery received came from "the back country" (i.e., the land that would become the base) and 54 percent of the co-op's sales of farm supplies ended up in the same area, in 1953 there was no practical purpose in continuing.[48] This was because the members of the co-op would be gone. And soon the productive woodlots and fields would be gone too.

Farm consolidation and outmigration has been a continuing feature of North American agriculture. The number of farmers in the army base area had declined by a third between 1931 and 1951. If Raymond and Lydia Scott had spent their married life strategically thwarting the ragged declension from "second nature" back towards "nature,"[49] reversing the retreat of the soils and timber on their home place from artefacts of agri*culture* and silva*culture* to wild lands (which happened during the decades of Harry Scott's inattention), their neighbours could see, as they travelled to town, that the holdings left behind were going back to bush and weedy waste. In the CFBC radio broadcasts, Eugene Morris, the Scotts' friend who managed the creamery, spoke about the "lost delight in things that have grown under our moulding ... the barns, stock, equipment and the land itself."[50] They were leaving neither wild land nor bush, but absent their presence they knew that wild land and bush would quickly reclaim the productive and congenial habitat that over five generations their dwelling had created. The place gave material form to their aspirations, and through a lifetime of practice they had embodied the place. For some, this was a connection so visceral that upon their being displaced from the valley, they too withered and died.

Two aspects of the built environment of the Nerepis Valley made the wrench of parting even more keen. The inhabitants of the lands taken into the army base did not live in isolated farmsteads but in ribbon villages, their dwellings and farm buildings within sight, scent, and hearing of one another along the Lawfield Road, their fields, pasture, and timber stands rising in common succession up the long lots behind. This landuse pattern facilitated the "changing works" among neighbours, which were so key to the enriching of the meadows and the culling of the woods,[51] and to the development of the dense network of community institutions that characterized the Nerepis Valley settlements.[52]

And in 1948, relatively early by comparison with most of Canada, rural electrification had come to the territory that would become the army base.

FIGURE 2.3   Lydia Scott and Frank Lacey at the Scotts' farm gate with cream truck, ribbon settlements in background | *Family Herald and Weekly Star*, n.d.

Because the late 1940s were times of good prices for both dairy products and farm wood, residents had seized the opportunity to improve their properties. Stables were equipped with electric milking machines, kitchens with electric cream separators, washers, dryers, and refrigerators. With electricity came water pumps and water heaters, which made it easier to maintain high standards of cleanliness in milking parlours and brought the pleasures of running water to homes. Recall that even with all this recent investment in electrical systems, equipment, and plumbing, five out of six householders surveyed at the time of the displacement had no debts.[53] Soon after acquiring these most bright and welcome modern conveniences for their homes and milking parlours, after four good years with a clear view of a prosperous modern future, the Scotts and their neighbours were required to drive away. They left expecting that, in their absence, their home places would go the way of the abandoned farms in their parishes.[54] Perhaps the war veterans among them had some foreboding of another future for the place. Perhaps they feared that once the farmer-lumberers and their families were gone, the soils of the meadows would be irremediably compacted by tanks and heavy equipment, their homes, barns, schools, churches, and community halls destroyed by artillery.

Three thousand people departed the base lands. Lydia Clarke Scott was forty-four and Raymond forty-eight in 1953, the year they left their home place at Summer Hill – too old, they decided, given they did not have

children, to make another farm. Raymond Scott told a Saint John reporter, "I'm all through. It took 17 years to build up what I've got, and I haven't
got the heart to start again. Life is too short."[55] Like most of those displaced
for the base, they relocated onto a property that was "quite old" and "required repairs."[56] Theirs was a partly renovated house on the edge of the
village of Gagetown. There they remained until they died, he thirty-eight
years later in 1991 at age eighty-six, she forty-three years later at age eighty-
seven. Financially, beginning again was not a hardship. Both had access
to non-farm employment. Lydia had completed normal school in Fredericton and had taught for four years at Hibernia before her marriage.
After they left the base lands she taught for four years at Upper Hampstead,
near the crossing from Long Island, where her ancestors had first settled.
There she led choirs in the Gagetown Music Festival and then taught for
twelve years at Upper Hampstead, each year upgrading her qualifications
at summer school until she was earning more than the principal for whom
she taught. Raymond was a heavy equipment operator who often worked
away on highway construction during their years on the land, and after
they moved to town, he eventually settled into similar work on the base.
He retired at age sixty-four and earned money in his later years through
his hobbies; she retired at age sixty, with twenty years of pension.[57]

But the move changed their marriage, their politics, their religious af-
filiations, their relationship to material possessions and their home, and
their role in their community. At Summer Hill, the Scotts had been "in
the forefront of new methods of doing things," actively involved in community as well as in agricultural organizations. He was respected as a
"terrific businessman," she as a "nice person who could dance, loved music"
and was a "good sport," "always doing something for fun."[58] They were a
handsome couple, posed in smiling embrace aboard their tractor for a
weekend glossy supplement distributed across the country in 1952, a representation of the physical strength and sensual vitality upon which the
hopes of postwar rural Canada depended.[59]

Lydia, a voracious reader and active student of public affairs, had from
1947 on kept scrapbook compilations of key events in her world. She was
the one to convene the Summer Hill meetings of the Farm Radio Forum,
nationally coordinated rural discussion groups animated by CBC radio
broadcasts on issues of the day. She served as a director of the Queens-
Sunbury Children's Aid Society, organized the blood donor clinics for
the Red Cross at Gagetown, and was active in the women's group at the
Summer Hill United Church and the auxiliary of the Orange Lodge at

FIGURE 2.4   Raymond and Lydia Scott on their tractor, with their vegetable garden, drive shed, house, and the neighbours' woodlot behind | *Star Weekly*, 22 November 1952

Dunn's Corner.[60] In addition to the co-op creamery, Raymond had organized a shipping club so that producers could sell in the competitive markets for pork in Saint John rather than be captive to dealers at the farm gate. He was a councillor for the parish of Hampstead on Queens County Council from 1946 to 1951, tax collector for the parish, and a politically active member of the Progressive Conservative Party who had contested local nominations for provincial office.[61] In April 1950, he was elected deputy grand master of the Loyal Orange Lodge (LOL). For the year 1951-52, Raymond Scott led the Orangemen of New Brunswick as grand master.

The LOL was a venerable vehicle for Protestant influence and Royalist enthusiasm, and a powerful instrument of political patronage. Grand master was a position of considerable moment in the public life of the province. The first Orange Lodge in New Brunswick was founded in Saint John in 1831; the lodge at Dunn's Corner near Summer Hill was LOL No. 4, begun soon after in 1836. Particularly in the Saint John Valley, through the mid-nineteenth century, the LOL grew rapidly. A careful demographic study concludes that nineteenth-century New Brunswick Orangemen were not preponderantly Irish Protestants but, rather, the descendants of New Englanders, Loyalists, and English immigrants, drawn together along these borderlands between the British colonies and the American Republic by anti-Catholic conviction and allegiance to British institutions. Twentieth-century Orangemen staunchly defended the symbols of the monarchy and the continuing significance of the Crown as a source of continuity and stability in the Dominion of Canada. The class composition of the leadership and membership was diverse.[62] Raymond Scott succeeded a physician as grand master. But his employment as a heavy equipment operator on provincial highways may have been germane both to his interest in the post and his electability, since in New Brunswick road work and patronage were near synonymous.

Historically in New Brunswick, the Catholics who were the focus of Orange attention and exclusion were Irish. The Catholic Acadians were in the distant north, their earlier presence in the Saint John River Valley a part of the forgotten history of the place (until the 1960s); the Catholic Irish were nearby, in Queens County along the Broad Road and at Petersville back from the Lawfield Road. On occasion, Protestant dominance and Catholic exclusion were enforced by violence, although this was not the case in the nineteenth century in Queens County,[63] nor, ordinarily, at LOL No 4. In the twentieth century, Catholics and Protestants along the Nerepis Valley often socialized on Sundays, for that was the day both groups climbed up to the meadows, on the forest verge where their properties abutted, to feed salt to and "visit with" their young cattle. Sharing in the twelfth of July celebrations of the Battle of the Boyne was another matter. As Lloyd McCann, a Petersville native, remarked dryly, "We had no white horses"; Connie Denby, a Catholic, remembers that the "old Orange Green hostilities generally came" to the surface at the annual Petersville community picnic "after a few drinks outside."[64]

The music and dancing of the lumber camps came back to the valley with the farmer-lumberers in spring. Musicians were people of standing in the community, their willingness to play an indispensable part of sociability along the Lawfield Road. Raymond was an enthusiastic fiddler – accounts usually dodge the question of quality[65] – and his group included Fred Francis on accordion and Gerald Wall on banjo.[66] Most dances took place under the auspices of the Orange Lodge at its No. 4 Hall at Dunn's Corner. There Catholics regularly danced. Certainly, Connie

FIGURE 2.5   Gerald Wall (banjo), Raymond Scott (fiddle), and Fred Francis (piano accordion) playing for a dance at Dunn's Corner, Loyal Orange Lodge No. 4 Hall | Public Archives of New Brunswick, P179/1

Denby's parents did, with her asleep on a bench until long after normal bedtime. But to play at the hall was akin to being host of the event. Should Wall, a Catholic, be permitted to play? Discussions were held. At Dunn's Corner Orange Lodge, the pleasure of the dance prevailed.[67] In the Nerepis Valley before 1953, when the Catholics present were English speakers of Irish descent rather than French-speaking Acadians, accommodations to facilitate shared pleasures were possible and readily made.

Scott would have just completed his year as grand master when the site for the Gagetown military training base was announced. The policy agenda for the Orange Lodge that year was unexceptional: preference for British and Nordic immigrants over all others; opposition to the appointment of a Canadian representative to the Vatican; satisfaction that, despite pressure on "the connection symbols between Canada and the Empire," "God Save the Queen" remained the national anthem and the Union Jack the national flag, and that royalist symbols and the British tie were being preserved.[68] As past grand master he presided over the last LOL No. 4 church parade in August 1952; by the glorious 12 July 1953, Summer Hill was a "ghost town." The similarity with the Acadian expulsion two hundred years before was left unspoken at the farewell service to mark the windup of the five Orange Lodges in the Camp Gagetown area. Instead, the presiding minister compared the "exodus of so many to the crossing of the Jordan by the Children of Israel."[69] But for the Scotts and many of their neighbours, the exodus from the Nerepis Valley fulfilled no promise. By 1960, most were less financially secure than they had been before 1952, on smaller holdings, cultivating less of their land, with more debts and a greater and unwelcome dependence on non-farm work. Their serviceable habits challenged by separation from a habitat viscerally known, many floundered and foundered, their present haunted by past vestiges whose material references had been removed or were in retreat.[70]

Once in Gagetown, a classic axially arrayed Loyalist village whose inhabitants regarded arriving backlanders with a studied hauteur, Lydia and Raymond Scott retreated from participation in all voluntary organizations except for their church. Without access to the LOL No. 4 Hall, Raymond, if he played the fiddle, did so privately for his visitors. His father had spent time in the provincial mental hospital, and Raymond, too, at mid-life, slid into a deep depression for two years, beginning in 1958. His second round of electroshock treatment cracked his back.[71]

Many said they left their hearts behind when they left the valley. Away from Summer Hill, Raymond seems to have lost the reckoning points that had guided the responsible, forward-thinking masculinity for which he had been known. He had said that he did not have the heart to begin a new farm. On the land and in the woods at Summer Hill, he had been confident of his judgment, of the reflexes and practices he had learned from the creatures and the contours of the place.

Part of this "lack of heart" may have arisen from the move itself. There was also the radical remaking of the landscape they had departed, changes evident to all nearby as smells and sounds emanating from the training area and reported in the press. Through two years of intense activity, soon after the Lawfield Road farmer-lumberers and their families left their land, the Nerepis Valley was remade as a component of Base Gagetown. At the peak, 2,200 woodsmen were employed in twelve separate projects on the site. In the fall of 1954, the goal was to clear more than ten thousand acres (4,046.9 hectares) of bushland. There was a queer unseasonable edge to the new activities in the once familiar place as work took place in summer rather than winter. Bulldozers uprooted trees; cranes hoisted root, trunk, and branch together "like matches," depositing them into piles. Huge bonfires lit the sky that summer. Come November, deep, heavy snow accumulated on ground that had not yet frozen; then woodswork had to be suspended lest the heavy equipment be mired in the mud. Slash, brush, and yarded piles were left behind in the retreat. In the spring of 1955, in a period of dry conditions and strong winds, fire raged through the regions back of Burton between Shirley Road and Hamilton Road.[72]

A new "second nature" was emerging in the woods. These timbered lands were props, interchangeable simulacra for Baltic and Bavarian forests, not legacies of local stewardship passed across generations. When clearing the timber (for weapons training) through mechanical means and fire could not keep pace with military requirements (remembering that these were lands known to provincial foresters for their rapid replacement rates, a boon to lumberers and a bane to those for whom the woods were in the way), in 1956 the army turned to contemporary chemical defoliants, including dioxins.[73]

And what of the meadows, judged by the director of the Royal Canadian Armoured Corps, on a January 1954 inspection visit, to be "tough, smooth, bushy and uncovered terrain" that was "ideal for tank training"? Early that June, when these fields were still wet from the spring floods, Sherman tanks, emerging "like giant steel elephants" from the practice mazes cut for them through the bush, "crunched through hedges, poked their blunt snouts into deep ditches," and sank deep. "This is the muddiest place I've seen since Treglio in Italy," remarked a sergeant-major from the Royal Canadian Dragoons. Wrestling for hours with the mud, diesel engines roaring "as though enraged at their own helplessness," the tanks made a din heard for kilometres outside the base.[74] The resultant compaction and the churning of layers of soil must have sounded like an audible death rattle to those who, through generations of shared practice and careful study of their particular seasonal variability, had nourished those meadows to agricultural productivity. These were lessons of a forgotten history, lost on the staff of the armoured corps who in their determination to find, back from Gagetown, an omnibus equivalent to Cold War terrains of European engagement were blind to specificities.

The clashes between familiar and unfamiliar were extreme: the sounds of mortar shells being launched over a nearby hill, the acrid fumes of ordnance and of hundreds of heavy vehicles performing manoeuvres along often-travelled roads leading to prospects once favoured for evening outings. The community hall filled with "sacks of Texas-grown carrots and crates of fresh milk" oddly packaged "in crisp paper cartons"; houses were made offices and barns became service depots sheltering Bren-gun carriers. Perhaps most darkly memorable was the mushroom cloud "blossoming" forty-five metres high and fifteen metres wide and resembling a "realistic nuclear explosion."

At least until August 1956, the whole camp area remained open to motorists so that those who wished could see the changed amid the persistent.[75] At the beginning of the 1955 deer season, Camp Gagetown was declared out of bounds to civilian hunters. Five local men were charged with trespassing and offences under the Game Act. Perhaps they were drawn back to their old haunts by newspaper coverage the previous summer of the fine sport being enjoyed by military members of the new Camp Gagetown Fish and Game Club.[76] Sometime thereafter, civilian visits were limited to the historical cemeteries on the base, allowed not on the anniversaries that might have moved mourners to their family plots but only on the Canadian military day of remembrance, November 11.

But some local people, among them many who had once lived there, went daily to work on the base. In the early 1960s, this civilian labour force numbered about eight hundred.[77] Local newspapers covered annual exercises at Camp Gagetown, the stories recounting battles between the forces of imaginary Blueland and Fantasia on the border of fictive Philistia, while the photographs showed soldiers marching along familiar bridges and "returning fire" as an aircraft swooped low over the steeple of a structure, which, unmistakably, had once been Summer Hill United Church. For several decades the built environment in the former Lawfield Road settlements persisted, useful to the military as landmarks and approximations of "real conditions;"[78] however, over time the weather, if not the weaponry, took its toll. In 1998, the remaining civilian structures were removed as hazards, leaving only the white post and page wire fenced cemeteries as reminders of the pre-1952 occupance of the site.[79]

In August 2003, on the fiftieth anniversary of the displacement, the Canadian Armed Forces opened the camp to returning families and provided bus tours of the territory that had become the base. In the absence of buildings, organizers had set out thirty-centimetre high black-lettered white signs identifying the location of stores, schools, community halls, and churches. Excitement was high aboard the buses as groups left the Gagetown fair grounds, but gradually the chatter subsided. Rounding Headline Ridge midway through the bright afternoon, one older man spoke, more loudly than he had intended perhaps, for his face soon reddened: "It's like driving around in the dark."[80]

FIGURE 2.6   Visitors to a cemetery on Base Gagetown at fiftieth anniversary homecoming, 2003 | Joy Parr

Little was recognizable of the past the displaced had returned to claim as the place they had made and where – through their work, laughter, worship, their accumulated knowledge of their spouses and traditions, their timber and their soil – they had fashioned themselves as sensing physical beings. Many of those who left in 1953 managed. Some were lost entirely. The Scotts' search for past recognizable vestiges took them on a shared journey that was by turns admirable and delightful, curious (or at least eccentric), and, in some respects – even to those who knew them well – utterly inscrutable.

At Summer Hill, the Scotts had been active in the church next door, originally part of an early nineteenth-century Methodist Circuit. The Canadian Methodists merged with the Congregationalists and some Presbyterians in 1925 as the United Church, a denomination that in west and central Canada, especially in urban areas, was closely identified with the social gospel. By the late 1950s, United Church ministers were also preaching liberal theology to more evangelical Maritime congregations. Lydia, an avid reader of the Bible and a conscientious parishioner, was dismayed. She and Raymond left the United Church at Gagetown, seeking a more theologically conservative alternative, and found it in the

Baptist Church in Upper Hampstead. This congregation became the centre of the Scotts' life in their later years. Lydia taught Sunday school, turning for inspiration to materials from radio Bible classes. After a dark interval, Raymond had a conversion experience. Theologically, as most Maritimers were moving closer to the 1960s Canadian mainstream, the Scotts reached rather to a part of the Clarkes' more fundamentalist past. Of their sizeable estate, a tenth went each to the Upper Hampstead Baptist Church, the People's Gospel Hour, the Radio Bible Class, and La Bonne Nouvelle, an evangelical group in nearby Moncton seeking to convert Acadian Catholics.[81]

"A sizeable estate" and a resolve "to convert Acadians": what have we here?

The Acadians, the French-speaking Catholics of New Brunswick, relatively minor figures in the histories of the Scotts and their neighbours while they lived along the Lawfield Road, were recast as central characters in the "moment of danger," which was the displacement. If in 1953 their loyalty to the Crown had made departing their land "unquestionable," even though they reasoned the choice of the location for the base to be mistaken, as the years passed their construction of these events changed. By 1960, those resettled indifferently on lands long out of cultivation or in places where good water was inaccessible began to think that the better course of action would have been to refuse to move. What had seemed a call on loyalty became, in retrospect, a trespass on rights, a violation "at least as traumatic as Evangeline," different from the eighteenth-century expulsion only in that the Acadians were driven out by "bad guys," while the base folk were expelled by "good guys."[82]

The rights revolution, as it came to Canada in the 1960s, was most successful in achieving near equal treatment for francophone Canadians, among them the Acadians. Both Canada and New Brunswick (uniquely among the provinces) became officially bilingual in 1969. Leaders of the political party with which the Orangemen of the Lower Saint John Valley were traditionally allied played key roles in effecting this change. In 1960, a Progressive Conservative prime minister, John Diefenbaker, enacted the first Canadian Bill of Rights in 1960; and in 1981, a Progressive Conservative premier of New Brunswick, Richard Hatfield, gave the provincial language rights institutional force by guaranteeing French and English access to distinct cultural, linguistic, and educational institutions.[83] Other changes happened under federal Liberal governments. The symbols of the British tie, which had been so important to Orange Loyalists of the

Nerepis Valley, ill accorded with the multicultural reality of postwar Canada. From the early 1950s, Orange protests notwithstanding, "Dominion" was dropped from the name of the country, the maple leaf replaced the Union Jack in 1965, and "O Canada" replaced "God Save the Queen" as the national anthem in 1980.[84]

The Scotts and many of their neighbours no longer recognized the bilingual, multicultural Canada of the 1980s. This was not the nation for whose strengthened defence in 1953 they had conceded the farms and woodlots they had made and had not wanted to leave to be unmade. In 1989, feeling disabused by both the "old line parties," Lydia became interested in the English Speakers Association of New Brunswick. That year both Scotts attended the founding convention of the Confederations of Regions Party in New Brunswick (CoR-NB), a party that opposed official bilingualism (especially as a basis for employment) and favoured the assimilation of the Acadians and immigration "pre-determined by the present ethnic mix of the nation" in order to maintain the dominance of the English charter group. CoR-NB had transient electoral success in 1991, particularly in the Lower Saint John Valley.[85] The appearance of bilingual signs on local roads, Acadian surnames in the provincial civil service directory, and public representatives with an Acadian cadence to their speech sorting out difficulties with licences, health care, and welfare unsettled many of the former base residents. Like Catholic Gerald Wall's playing at the LOL No. 4 Hall, bilingualism widened the definition of who might occupy the stage and play host in the province. But whereas acknowledgment of the shared bodily pleasures of music and dance was enough to forge consent to Wall's presence on the podium among neighbours along the Nerepis Valley, official recognition of the Acadian claim to equal presence in the public institutions of New Brunswick entailed a thorny parity. Could the expulsion of 1758-59 from the meadows and woods around Grimross and the displacement in 1953 from the meadows and woods back from Gagetown (both implying a changing North Atlantic military order) – expulsions that concerned the same ground being made from nature to productive second nature by similar skills applied in pursuit of similar goals – be recognized as parts of a shared history? Maybe sometime, but not in the Scotts' time.

Perhaps there was something distinctive in this case of unsettlement, the early dawning conviction that the nation arising on the foundations of their sacrifice was not the nation they had hoped to sustain. More likely, as may become clear in the case studies that follow (of the villagers along the St. Lawrence, the shepherds and cottagers along the Huron Shore, and the ranchers and market gardeners along the Columbia River), forcible unsettlement from habitats that are viscerally known is as grave a calamity for humans as it is for other adaptive living beings. Habits being held deep, habituation to habitat being beyond telling, the implications of unsettlement are beyond consciousness, eluding planful preparation, always indwelling, threatening to rise up to haunt one in "moments of danger." Volition and scale seem to be key matters in the remaking of historical bodies so that they may thrive before new sensory challenges, and the nuclear safety and safe water narratives that are to come are happier stories.

But before we move on, let me tell the story of the Scotts' sizeable estate, a playful coda to Lydia's playfully lived life and a testament to her fortitude, her love for Raymond, and her capacity for forgiveness. The million dollars ($982,460.78 to be exact) distributed among the nieces and nephews she and Raymond left behind, to her brother Willard, and to the Canada Heart Fund as well as to the gospel institutions that had become close to her heart, came, in part, from the Scotts' investments. The Scotts were notoriously parsimonious, and the 1950s and 1960s were good to Canadian investors. However, throughout the 1980s, they had another source of income. Lydia had helped Raymond overcome the depression that had gripped him in the late 1950s by urging him to hook mats, a craft he had learned from his mother. Raymond was colour-blind, and by all indications form-blind as well, but Lydia loved colour and drawing, and she enjoyed teaching art to the children of Gagetown. Lydia's designs included the conventional portraits of their farm at Summer Hill, her father's mill at Hibernia, oxen drawing a plough, and the truck that took their cream to Hampstead. There were many, many portraits of pigs (usually portrayed as portly and kissing), kept to fatten on the milk left after the separation of the cream. Willard Clarke estimates that there were four thousand mats in all, completed at an average of two a week by an increasingly obsessive tobacco-chewing Raymond, fiercely hooking eight hours a day in their garage and attic. The unconventional parts of the oeuvre are portraits of naked women. One rises from a pool before a stand of birch trees. Another has set her book and her horn-rimmed glasses aside (Lydia wore horn-rimmed glasses) to pose seductively for the hooker. Today, some of the rugs are in Canadian museums and the private collections of wealthy New Brunswick families; however, throughout the "folk boom" of the 1980s, an astute intermediary placed over a thousand of them with American and other international collectors at very good prices. A book about their work, *The Gagetown Hookers*, written by their dealer, Larry Dubord, shows them on the cover, smiling above mats of happy kissing pigs.[86]

http://
megaprojects.uwo.ca/nuclear

This site features the new media work associated with Canadian nuclear operating procedures and training at the Bruce Nuclear Power Facility. It consists of sound compositions exploring the unique training philosophies, regimens, and work procedures of several of those people associated with the Bruce nuclear power facility. The sound essays feature excerpts of interviews conducted by Joy Parr with those who lived near, worked in, or administered the nuclear facility.

The sound essays for the nuclear training portion of the Megaprojects New Media series are well founded and inspired by prior work, notably Glenn Gould's *Solitude Trilogy* radio documentaries. They bring into focus the interviewees' voiced self-representations of having learned, taught, and elided the epistemological and embodied demands of the dangers of radiation.[1] The two sound essays employ select passages, repetitions, and elisions.

"The McConnell System" explores Lorne McConnell's radiation protection philosophy: all workers, from plant managers to janitors, would have knowledge about the safety measures for which they were responsible.

"Equate Things" illustrates how in passing on radiation and hydrogen sulphide gas protection knowledge to plant workers, the early nuclear administrators and trainers resorted to familiar farm- or factory-honed analogies that would resonate well with the local labour force.

—JV

3

Working Knowledge of the Insensible: Radiation Protection in Nuclear Power Plants, 1962-92

Radiation is a workplace hazard that eludes the sensing body, or seems to. After Chernobyl and Three Mile Island, Kai Erickson, a student of hazards and technologies, described radiation as "an invisible threat," "the very embodiment of stealth and treachery."[2] The first generation of Canadian nuclear power workers, from their four decades of experience around reactors, have a different sense of the matter. They describe a physical awareness of radiation, its morphology and topography, and a training transition that schooled them to embody the symbolic readings on their instruments as sensation and to use those measurements to inform their actions as they produced power. Just as the people displaced for Base Gagetown came to know viscerally through observation and systematic technical study the meadows and woods whose new natures they remade by their labour, so the men and women who worked within the spotless, silent spaces of 1960s and 1970s nuclear generating stations formed intimate connections with the reactors and their radiation fields. By "listening to its very cries," they developed a "feel and a touch for the plant." Though the theoretical studies of ionising radiation that nuclear workers needed to master in order to maintain production and keep themselves safe in the generating station were more arcane than the agronomy and silviculture passed on by elders and agricultural representatives along the Lawfield Road, the stories of Gagetown farmers and nuclear workers share an emphasis on attentiveness and alert expectation. This tuning of body, technology, and environment was, in one case, to

seasonal changes in meadows and the woods and, in the other, to meaningful differences among engineered regularities in the systems of the
station. By their telling, "doesn't feel right" ceased to be a metaphor among
nuclear workers about their workplace circumstance and, through study
and practice, became a report from their sensing bodies[3] – they appraised
the radiation fields in which they laboured, like the fields of oats their
siblings and forebears had tended, by reference to sensory histories of
practice and perception.

From this perspective, perception is not only an encounter between a
sensing body and a discrete thing but also a "way of being in the world,"
transcending the boundary between body and object. Canadian radiation
workers likely did (they said they did) embody knowledge of their task-
scape, of the physical structures of the generating station. They also developed ways of attending to the insensible radiation fields of their
worksite, somatic modes that kept them safe, sensibilities and practices
that were refined as attributes of Canadian nuclear work culture. This
chapter is about the experience of teaching, learning, and practising
radiation protection, and the distinctive theoretical and practical training
and work organization that prepared workers to embody the insensible.

The Canadian nuclear safety culture differs from those around nuclear
reactors at Savannah River in the United States and at La Hague in France.[4]
How nuclear workers embodied the insensible depended both upon the
design of the reactor in which they laboured and the balance between
hierarchy and autonomy culturally understood to best mitigate risk. As
Gabriella Hecht has shown for France and Itty Abraham for India,[5] nuclear choices were influenced by national, political, and managerial cultures.

Canadian nuclear reactors are technologically distinctive in several respects. They are fuelled by natural rather than enriched uranium. They
are moderated and cooled by heavy water, deuterium, rather than light
water (the production of heavy water is discussed in detail in Chapter 6),
hence the CANDU name, CANada Deuterium Uranium. The system is
pressure-tubed. The fuel bundles are set horizontally so that the reactor
can operate continuously without shutdowns for refuelling. These technical differences influence the physical structure of CANDU reactors and
the topography and composition of their radiation fields. And they contribute to the differences between Canadian radiation protection practices
and those in the United States and France.

The focus here is on three Canadian nuclear installations. The twenty-
two-megawatt nuclear power demonstration (NPD) plant was the first

Canadian reactor to produce power, a pilot built on the Ottawa River (near the Chalk River) two hours' drive north of Ottawa. The NPD went critical in 1962. Here, the radiation protection practices characteristic of research reactors producing experimental results, plutonium, and medical isotopes were radically reorganized to suit the new project of generating electricity. The 206-megawatt Douglas Point reactor, located on the eastern shore of Lake Huron near Kincardine, Ontario, was the first Canadian reactor to supply commercial power to the grid. There, from 1968 to 1984, distinctive Canadian radiation protection protocols were implemented and refined. Douglas Point was a veritable festival of "teachable moments" as power workers were simultaneously struggling with flawed pressure tube design and the unhappy discovery of tritium, an ambient beta source that was created as heavy water passed through neutron fields. Point Lepreau, a New Brunswick Power CANDU 6 reactor with the capacity to produce 634 megawatts, began operation in 1982 on the eastern coast of the province between Saint John and the Maine border. Here, the distinctive radiation protection pedagogy and practice worked out through the 1970s as the eight Ontario Hydro-built five-hundred-to-seven-hundred-megawatt reactors at Pickering and Bruce were refined.

The early Canadian experimental reactors were located in universities and two Crown corporations, Defence Industries Limited (DIL) and Atomic Energy Canada Limited (AECL). Radiation protection in these federal agencies, from whence many pioneers came to the nuclear power industry, was the responsibility of health surveyors. The health surveyor inspected the worksite and wrote out a permit specifying the clothing, shielding, and dosimetry required and the allowable length of exposure. Other early nuclear workers had little radiation protection training. When he started at Chalk River in the early 1950s, Charles Mann remembers being given four pages of typed onion-skin paper as his manual. The rest involved learning on the job and learning by word of mouth. Gerald Black recalls receiving a day and a half of instruction when he began work with AECL at Chalk River in 1958. Workers generally didn't know much about radiation and were obliged to depend on the surveyors as "shepherds" of their work in radiation fields. This hierarchical system had advantages in experimental sites, where the work was varied and presented a wide range of radiation hazards.[6] After the transition to power production, these radiation work rules prevailed in the nuclear generating stations of France and the United States. Experimental reactors apart, they did not persist in Canada. The French and American work rules required employees to be less autonomous and more deferential to authority.

The hierarchical system produced tensions. A shepherd tended sheep. A mechanic, electrician, or millwright with a task before him might have to wait around for hours until a surveyor was free to plan the job. Ken Hill, by the late 1980s a widely experienced nuclear operator long certified to be responsible for his own radiation protection, remembers with exasperation the callow air force officers, "real snot bags," surveying his work at AECL research reactors near the end of his career. Engineers and tradespeople alike resisted the surveyors' authority and suspected their competence. Without independent grounds to assess radiation hazards, workers were tempted to "find easier ways" following the conventions of their own trades, taking risks "without realizing the nature of the risks they were taking."[7]

Two incidents from these early days have endured in Canadian nuclear history as forbidding folklore and telling instance. On 12 December 1952, during a routine shutdown, a National Research experimental reactor, the NRX, melted down. Charles Mann, who was present and participated in the ensuring fourteen-month cleanup, recalled, "It literally melted itself into a glob." The AECL official historian links the cascade of misheard and misinterpreted instructions from supervisor to operator, which created the event, to the agency's "rigidly hierarchical system of authority and responsibility."[8] The second incident involved one man, Alex Sandula, a mechanic who, unawares, picked up a "rabbit," a piece of fuel rod that having passed though the reactor neutron flux then ricocheted off the shielding and onto the floor. In the moments before he reached a monitor, heard the alarm, and dropped the rabbit, contamination measured at three thousand rad was deposited in the creases of his fingers. For years his arm erupted in sores. He eventually lost his hand.[9]

Radiation is a protean threat, and workers who knew little about the radiation hazard had no sound information with which to conceptualize the danger in the job site; to feature the geography, the topography, and the heterogeneous intensity of the fields; and to embody the hazard in reflex and intuition. If they did not know that alpha, beta, and gamma rays and neutrons spread in different ways, had different penetrating capacities, and would be produced in different parts of the plant, then the hazard of the nuclear job site was not only insensible but also inscrutable. The shop-floor oral culture of the early years, the emulation of senior men, the four pages of instruction, and the few hours of training taught workers something about what to do in the presence of radiation fields, but it taught them nothing about why. In the view of John Wilkinson, whose nuclear savvy dated from his wartime activities with the Scientific and Technical Intelligence Branch of the British secret service, "You couldn't carry on this little apprenticeship system with a whole station full of people."[10]

Though the CANDU power plants would present a narrower spectrum of radiation hazards and a more stable compass of tasks than early experimental piles in the universities and at Chalk River, the power plants would be industrial workplaces with more people of more varied backgrounds in and around the reactor building.

The man in charge of the NPD pilot to nuclear power was Lorne McConnell, a University of Saskatchewan-trained physicist who was put in charge of the NRX emergency soon after the event and then organized the cleanup. At NPD, McConnell instituted what seemed, though he himself denies this, a radical reorganization of radiation protection practice, innovations designed to make occupational health and safety in nuclear job sites analogous with those in conventional process industries. Every person in a station producing nuclear power was to be responsible for her or his own safety. Each worker from janitor to senior operator and shift supervisor would be prepared at least to the level of the health surveyors in the AECL system, able in the course of their job "to take their own contamination readings, scan themselves, look for radiation sources and safely operate the equipment."[11] Each was to know the procedures and to understand the plant sufficiently to make sense of them. The early 1960s aspiration was that nuclear power workers would thus be equipped to make reasoned decisions, "to take risks by changing the way they were doing things ... knowingly, in the same way that people would be taking chances with conventional hazards," to "look after themselves implicitly" and to look out for the safety of their co-workers.[12] The changes in radiation protection were paralleled by a shift away from specialization in job classes, so that mechanical and technical maintainers developed a wider working knowledge of the station and were safely at home with more of its elements. "Safe as at home" was McConnell's motto for the change.[13]

Between the plan, which had the potential to significantly extend the embodied working knowledge of every person who entered radiation fields, and its execution lay conceptual and pedagogical challenges. Because at AECL the health surveyors had been trained through an apprenticeship system,[14] Charlie Mann's four pages of onion-skin paper apart, there were no texts, no symbolic representations of what surveyors knew sensually and what they did in practice. And so a drafting team, consisting of the newly appointed NPD health physicist; his assistant, Robert Wilson; two health surveyors seasoned in the AECL hierarchical system, John Wilkinson and Frank Caiger-Watson; and the assistant superintendent of the NPD, began to collect the oral culture and habits of the surveyors to combine these with the physics and chemistry the scientists judged relevant, and to formulate procedures. Along the way, the team consulted documentation from the American research establishments at Hanford Ridge and Oak Ridge. Then, over a period of months, McConnell cajoled the group to reframe what it knew from its own and US hierarchical precedents into procedures a high school leaver responsible for his own radiation protection could follow while performing his job.[15]

The commissioning and operating manuals for the pilot power station, the NPD, were similarly derived from practice: "When we figured the system was working, 'it's commissioned,' we had to write it up. That was the commissioning manual." Operators studied the manufacturers' documentation and the plant flow sheets, and then got the equipment functioning. Their reports of what they'd done were the early operating manuals: "You put down your best guess ... put it down in an operating format." While their writers thought of these texts as "basically a history," the history was short. These were not chronicles of proven routines. And so, at the NPD, when working knowledge was evolving rapidly, there "wasn't really ... a hard and fast rule that you had to follow the manual." "You were allowed to think, how do I fix this thing?" Only later did deviation from the manual become "sacrilege."[16]

The labour force of the first Canadian reactor to produce power was drawn from the hydraulic and thermal power divisions of Ontario Hydro and from the AECL research establishment at Chalk River.[17] These "dam attendants," "coal and ash handlers," and "nuclear jockeys," as they called one another, were all entering new shop-floor terrain. None had previously been responsible for his own radiation protection. None had worked on a site where mastering and maintaining documentation was so central to the job. The men from the experimental reactors had a grasp of nuclear theory and "a feel for working in radiation fields." The thermal men were familiar with the workings of the NPD steam-driven turbine hall, because outside the reactor building, a CANDU nuclear plant is essentially a steam-generating station.

Though the contrast can easily be overdrawn, the men from conventional power stations were leaving worksites of roiling water or roaring boilers, where workers had physical access to all parts of the station at all times, essentially steam plants where pulverized coal dust hung in the air about open control rooms. These were taskscapes suffused with sensory stimuli. Jim Bayes, who came to the NPD reactor after four years as a stationary engineer in the R.L. Hearn steam plant in Toronto, described the nuclear working environment they entered as mildly and minimally different, "a little more sneaky and a little more subtle." These were small but notable. Many men who by trade had tuned their bodies to register the mechanical and hydraulic systems around them found the sensual banality of the nuclear site "hard" to accommodate. This is how Ken Hill, a refinery worker, described the change after he entered the nuclear industry in 1970:

> In the oil industry, you know sound is there ... sound is more important. If the sound changes I've got to go there and find out why the sound changed. Pumps shutting down or pumps starting or – rrrrrrr – like that. Whereas in a nuclear plant you can walk by and a pump starts, a pump shuts down and it could not mean very much, a guy's changing-over pump. Tank level got down so far and that's it. I found that kind of hard because I was listening – coming from the oil refinery – it makes a difference.

Bob Ivings, who came to the NPD from a coal-fired plant, remembers, "When I was an operator at Hearn, one of my biggest assets was my nose. Because if things began to overheat and you had a bad bearing or something, you were walking around and you'll smell it."

In terms of responsibility, new radiation protection protocols at the NPD made safety in nuclear power stations more analogous with those in conventional process industries. In terms of sensuality, distinctions between nuclear and conventional industries remained. The difference is significant. For the new protocols to be most successful, nuclear workers would need to retune the embodied practice and working knowledge they had used in conventional worksites. They would need to cultivate a different somatic mode of attention.[18]

In retrospect, nuclear workers who worked at Bruce and Point Lepreau recalled welcoming the new system and the individual responsibility it conferred. Those who stayed with the work and adapted to the change took on this competence to keep safe, in the presence of a grave insensible

danger, as a personal attribute: "I preferred to look after my own self."
Claiming the workplace convention as consistent with the concept of
being self-made was key. Knowledge embodied only by repetition in accordance with obedience to authority would have been held rigidly.
Knowledge volitionally embodied was held reflexively, within the stance
of alert somatic attention through which the worker had deliberatively
accepted the re-embodiment. This was a collegial rather than a military
model of workplace relations.

At the start, in 1962, the AECL and the Canadian regulatory agency,
the Atomic Energy Control Board (AECB), were sceptical about the initiative. Nuclear power producers in France and the United States came to
look around but, after sizing up the formidable front-end training costs,
turned away.[19] The system that made workers responsible for their own
radiation protection was implemented in the nuclear stations of Ontario,
Quebec, and New Brunswick and was exported with CANDU reactors
to India, Romania, and (later) China. In the years before 1983, the Canadian system of radiation protection was the exception to the prevailing
health surveyor model at nuclear sites around the world.

The two systems rested on different epistemological foundations. For
example, in the 1980s, Dupont managers at the Savannah River project in
Georgia, refusing the premise of the NPD innovation, insisted that industrial safety and nuclear safety were not equivalents.[20] The American
approach involved prescription: "codes you had to comply with," "procedures ... for gosh sakes don't ever deviate from them," even to employ
a safer alternative. Contrary to this, in the 1970s the AECB, then the
Canadian regulator, had come around to the position that "as you could
never write a perfect procedure," deliberative refinements made by well-
trained and long-practised workers were acceptable alternatives to code:
"Just show us it's safe."[21]

These dissimilar judgments about safety, autonomy, and codified and
evolving working knowledge were rooted in history. In the United States,
most nuclear managers and engineers came out of the navy, for Americans
had at first been interested in power reactors as a way to drive boats. The
hierarchical culture of defence establishments carried over into nuclear
generating stations during the Cold War and thus into managerial decisions about radiation protection.[22] The Americans had favoured light over
heavy water nuclear technologies in part because they were better suited
to submarines.

The refuelling of light water reactors required a major shutdown. For
this work and for major maintenance, American producers of nuclear

power relied upon a mobile contract labour force that travelled about performing these specialized tasks in reactors of varying designs. Neither contractors nor power companies were willing to make a prodigious investment in radiation protection training for these workers, training that would have had to be specific to each plant design. Rather, the health surveyors familiar with the facility who shepherded the smaller staff at power had their busy season guiding the contractors during shutdown.[23]

CANDU reactors dating from the NRU, the immediate successor of the NRX, were refuelled at power. Deep investments in training their stable year-round labour forces made sense to Canadian nuclear operators. In CANDU reactors, anyone who worked in radiation zones was also qualified to supervise the radiation protection of others, including the smaller number of outside contractors who entered the station. This intensified surveillance. If a contractor embarked upon something foolish, "somebody would say, 'don't do that.' It might be a janitor, it wasn't somebody specially paid for that."[24]

The CANDU refuelling technology allowed power plant operators to provide continuous employment and led them to emphasize skills development among local recruits, and public policy promoted this human resources strategy. The nuclear stations at Bruce in Ontario and Point Lepreau in New Brunswick were sited in marginal economic zones at least in part so that they could be generators of stable high-waged employment.[25]

Nineteen hundred and sixty-two was a good year to implement a fundamental reorganization in work rules in Canada, for at that time, Canadian power producers and their unions had only a few years' experience using steam to generate electricity. Hydraulic sites had predominated (electricity is still commonly referred to in Canada as "hydro"). There was relatively little inertia to overcome when implementing new work rules in steam generating stations, where the heat came from nuclear reactions rather than from burning fossil fuels. The power workers unions were small. By then, similarly propitious times had perhaps passed in both the United States and France. Steam generation had a longer history in these jurisdictions, including a union legacy of job classifications and working knowledge refined through bargaining to organize tasks around boilers, turbines, and transformers.[26]

By the late 1980s, a finding key to inquiries about working embodied knowledge was emerging from comparisons with hierarchically organized nuclear worksites. Workers with narrower responsibility and accountability were more likely bored and thus less attentive. Workers made more accountable for their own actions and for those around them were more "alert and aware of themselves and their surroundings," more responsive to little cues and changing conditions, likely to "notice and confront discrepancies earlier." A person given greater responsibility and account-ability assumed a different somatic mode of attention. As the authors of a 1989 study reported, "They can almost 'sense' when a particular piece of equipment is not working right, before the problem is big enough to be a major problem."[27] These workers were more likely to use the know-ledge they thus embodied to optimize performance and safety.[28]

Françoise Zonabend's findings from the La Hague nuclear installation in France are similar. Ex-employees protested, "The work's 90 per cent routine with no say of your own." In France, as in the United States, workers were vexed by their client relationship with members of the Radiation Protection Department, "part guardian angels, part police." Utility representatives noted that surveillance removed an employee's "motivation ... undermined his perception of reality." Workers taking their own radiation readings and analyzing contamination levels in their own samples made guides for action from numerical representations of insens-ible hazards. They organized their work accordingly,[29] and they were more likely to work safely.

The work safety challenges at Douglas Point, the first Canadian re-actor to produce power for the grid, were memorable and instructive. Scale alone altered workers' sensory modes of attention. Douglas Point, at 200 megawatts, produced ten times more power than the NPD and was phys-ically ten times as big. Whereas at the NPD a chief operator could see half the station from the control room window, at Douglas Point "a guy would leave the control room, you had no idea where he was." Operators were moving from visual to instrument surveillance in unpropitious circum-stances. Planning for the transition from pilot to power had proceeded deliberatively through the recession years, from 1958 to 1962. But in 1963, as demand for electricity increased, the scientists designing Douglas Point, gatekeepers of a new source of supply, were pressed to complete their work. Their haste showed. Douglas Point was "delicate." What, in McConnell's words, should have been designed "like a dray-horse, was designed like a race-horse," a reactor made by scientists that did not satisfy engineers.[30]

The high pressure tubing in the heat transfer system quickly sprung leaks of tritiated heavy water. The valves imported from the utility's conven-tional steam plants shed rare and radiated heavy water at rates intolerable even if it had been tap water. At least 25 percent of station radiation doses came from moderator and heat transport system tritium leaks. Beta radiation levels were sufficiently high that for tasks in the reactor building workers had to don plastic suits daily rather than exceptionally. Stellite from the pump seals borne in heavy water through the reactor yielded cobalt 60, an active beta-gamma emitter. Upkeep at Douglas Point exceeded projections by 600 percent.[31]

A "primitive, early computer" measured the flow and temperature of the plant's 308 coolant channels. The computer and all the plant instrumentation was set within a narrow – in retrospect a too narrow – band. An imbalance in the readings signalled the computer to shut down the reactor. Reactor trips were frequent, sometimes as many as three per shift, a safety feature of ambiguous implication since the reactor was most vulnerable on the way down from and back to power. Work in the Douglas Point control room was so nerve-wracking that some experienced operators transferred to less daunting jobs. Getting the work done within the combined yearly radiation dose limits of the station's staff complement was a near thing.[32]

Nuclear workers of this period are legendary. Given the prevailing knowledge and equipment, they had to be innovative. As innovators they assumed a distinctive "pioneer" risk calculus: "Nobody in the world knew what we were doing then, including us ... Basically, there was a lot of theoretical stuff but nobody had ever tried it out."

Effective radiation protection practice is an arbitrage of exposure among time, distance, and shielding. In the Douglas Point days, when their equipment was rudimentary and flawed and their knowledge of their hazard was gestational rather than resolved, nuclear workers emphasized speed: "You quickly learned that rather than tippy toe around, quick in and quick out resulted in less dose, plus gets the job done." Pioneers, who had participated in the NRX cleanup, who had waded through heavy water in Chalk River emergencies, or who had been the first in radiation fields before there was much knowledge about the hazard, assumed all radiation knowledge would be imperfect. They thought of themselves as "task-oriented" and were more likely than younger men to act rather than to deliberate.[33] In the French industry, men with this orientation towards risk, emphasizing speed and efficiency, were called *kamikazes* and were distinguished from *rentiers*, workers who managed their radiation dose allowance as thriftily as a person living on modest property revenues. *Kamikazes* more often worked in the mechanical part and *rentiers* in the chemical part of a French generating station.[34]

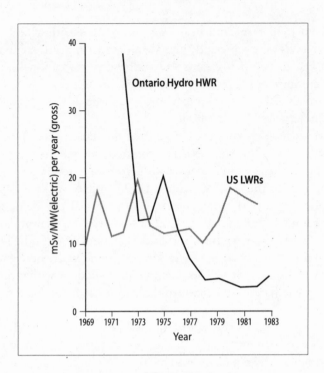

FIGURE 3.1   Collective dose per unit of electrical production, 1969-83. | *Source:* Adapted
from original in Robert Wilson et al., "Occupational Dose Reduction Experience in Ontario Hydro
Nuclear Power Stations," *Nuclear Technology* 72 (1986): 243-44.

In Canada, the distinction seems to have been generational. The Canadian novelist Catherine Bush, whose father worked the Canadian nuclear industry, has a character parse the different responses to radiation dose
exposures in this way: "Nuclear cowboys ... were reckless, cavalier in the
face of radiation, believing they could do anything ... nuclear fishermen
were careful and approached their source of power as if it were a fruitful
but dangerous sea: they took only what they needed and respected its
dangers."[35]
The training the pioneers got at the NPD and at Douglas Point was
more time consuming and more theoretical than the training they received
at their previous jobs. At the NPD and Douglas Point, the classroom
radiation protection training alone was full time for six or seven weeks. In
health-surveyor systems in France, workers typically took two days of
theoretical and two days of practical training in radiation protection.[36]

In the 1960s, the Ontario Hydro course included the physics, chemistry, and thermodynamics of radiation, radionuclides and radiation protection, "the basics of radiation external exposure, how to control internal exposure ... some ideas of the biological effects of ionising radiations,"[37] all of which were essential foundations for the working knowledge that would allow employees to get their jobs done within their yearly dose limit. When Ontario Hydro declined to pay nuclear workers a premium for their radiation protection knowledge, their union took the issue to arbitration and, based on their "special skills and responsibilities," won for nuclear workers the equivalent of the highest local cost-of-living allowance in the contract.[38]

It is hard not to sympathize. Science aside, even the calculations required of workers estimating dose per task was a considerable challenge. Local hires for Douglas Point were workshop savvy high school leavers from surrounding farms and furniture factories.[39] At Point Lepreau in the late 1970s, where New Brunswick Power made local hiring a priority, some recruits had been fishers and woodworkers. Others from skilled trades, for example machinists from the nearby Saint John shipyard, were enticed by the prospect of joining a leading-edge high-technology industry. Many had been away from the classroom for a decade or more, and many had left school after Grade 8.[40]

Robert Wilson's teaching for the expanding Ontario Hydro nuclear program in the 1970s, and Jan Burnham's pedagogy at New Brunswick Power in the 1980s, took on this next challenge. Their task was to make nuclear workers with embodied working knowledge to manage their own safety in radiation fields, not to choose experienced experimental defence or power workers but, rather, to carefully select and train workers from the local labour pool. Burnham's radiation protection manuals, in their five editions, became the standard texts for the Canadian system of radiation protection, the basis for embodied working knowledge of radiation fields, and the foundation for the somatic modes of attention that guided CANDU workers.[41]

With an English master of science degree in nuclear instrumentation, Burnham was thirty-four in 1974, when he transferred from Ontario Hydro to New Brunswick Power to establish health physics at Point Lepreau. A gruff northerner of hard-living postures and close, sympathetic habits of observation, he began well, showing respect for the working knowledge new hires brought to the station. These were "very smart guys, they could fix anything." He gave his manuals the format of high school science texts, which was fitting since men worked on them in the evenings next to their teenagers who were doing homework around the kitchen table. New workers at Lepreau, with gaps in their book learning, took a week's preparatory instruction in mathematics. Those for whom math came easily unobtrusively continued study sessions for those who were struggling. After all, he affirmed, "this stuff isn't rocket science, right?"[42]

The foundations for a working knowledge of radiation fields and for the somatic modes of attention appropriate to this insensible hazard were theoretical. The 1979 manual opened with chapters on the atom and radiation theory since the various types of radiation have different physiological effects and need be differently embodied. With this knowledge, recruits began to reconceptualize their working bodies.

Consider these examples, woefully simplified though they are. Alpha particles move quickly over short distances. Externally, they pose little threat since they will not penetrate the dead outer layers of skin. When ingested or inhaled, however, they are a danger from which movement provides no defence. Alpha particles come from fuel particles tramped out from the fuel bays and fuel-handling areas. Somatically, Burnham quipped, these are hazards "you can't walk away from."[43] Beta particles penetrate about a centimetre of human tissue and present in a wide range of energies, some sufficiently strong to burn the skin. The body itself is thus an ineffective shield against beta. A sheet of plywood or rubber sheeting will serve. Between body and beta, lead blankets are the shielding of choice. For gamma radiation, evasive action is never completely successful. Water, aluminum, and concrete barriers make measurable dents in gamma intensity but some gamma will penetrate all shielding. Gamma is a radiation hazard that nuclear workers manage on a bodily uptake budget year by year, attempting to keep their accumulation as low as reasonably achievable (ALARA). The problems neutrons present to bodies differ from those presented by gammas. Both neutrons and gammas are produced prodigiously in the reactor core at power, but neutrons activate matter: they create radioactive progeny, including progeny made with the water that constitutes much of the human body. From heavy water, they create the hydrogen isotope tritium, the beta hazard first formidably encountered in CANDU stations at leaky Douglas Point. Tritium is an airborne hazard, which, once inhaled or ingested, joins the body's fluids. It is sweated and excreted, but it also fractionally persists.

Time is a factor in these bodily encounters. Dose accumulates with time in a radiation field. Some fields pass away. Most gamma emitters decay quickly; during shutdowns, the gamma hazard soon dissipates. Others do not. Tritium, the beta emitter, has a half life of 12.3 years. Cobalt 60, a beta-gamma emitter that plates out from many steels and that tramps about as heavy water carries it though the tubes of the plant, has a half life of 5.3 years. Waiting won't help.

What is to be somatically attended to also depends upon place. Neutrons are contained behind the concrete walls of the reactor core. But the moderator system (the heavy water that slows neutrons in the reactor to cause fission) and the primary heat transport system (the heavy water that circulates from the reactor core to the boilers generating steam) extend beyond the protecting shield of the reactor. Where tritium leaks from the moderator pipes, the bodies of workers nearby meet beta. Where particles of cobalt 60 have ceased tramping and have come to rest along the pressure tubing, nearby bodies meet beta-gamma.

This rudimentary knowledge that alpha, beta, and gamma rays and neutrons spread in different ways, have different penetrating capacities, and are present at different parts of the site suggests how the insensible hazard can become both sensible and scrutable. In real life, this knowledge, which was elaborated, complicated, and refined over weeks of training at Point Lepreau and refreshed at three yearly intervals, prepared nuclear workers to elude or minimize the bodily effects of radiation. Their working knowledge was a skilled arbitrage of time, distance, and shielding, a performed repertoire of embodied practices guided by instruments.

Burnham's academic specialty was nuclear instrumentation. Some of the radiation protection devices at Point Lepreau were of his design. His fascination with the inner workings of measuring instruments shows in the early teaching manuals. This attention exasperated students. For them, radiation measures were novel and a conceptual challenge. In the mechanical and process industries from which many had come, instruments were less likely to be black boxes than devices whose workings were apparent: "You could see the little plunger going up and down." Or whose implications were accessible by analogy: "As the level in the vessel rose, so did the level on the gauge." In a nuclear station, instruments were electrical. They yielded readings, symbolic representations of the insensible radiation field. These readings did not direct action. They were inputs for calculations and analysis of whether a job could be accomplished within acceptable dose limits. Workers were not so concerned about how the instrument worked as about how to become quick and competent with the calculations that defined the insensible in their specific tasksite.

Zonabend observed that French nuclear workers, who carried meters but did not do their own dose calculations, treated the instruments as *means of protection* rather than as measures of radioactivity. Their own dose ciphering spared Canadians workers this mistake. Still, edition by edition, power workers persuaded Burnham to "waste less" of their time mastering the electronics of instrumentation and to cut to the chase, to the analytical work that let them "see" the dimensions of the hazard before them.[44]

Their differing innards made instruments read changes in the radiation field with different response times. Some had dead times between registering pulses. To this extent at least, Burnham was right and his students were wrong. Workers had to learn to embody these affordances in their practice, to adapt their gait to travel within the registering pace of the device and their muscles to position the instrument for the most accurate reading, to choose an instrument with the best trade-off between response time and accuracy for the job. Through the 1980s, response times diminished, especially for low doses. But workers still recognized the relationship between bodily gesture and good information, particularly in the "friskers" they used at boundary points in the station to measure gamma contamination on their persons. The applications section of the training course and the year of probation that followed it were an apprenticeship, when workers reschooled their reflexes to embody the symbolic reading on the instrument as a sensation, as a form of tacit knowledge with the same credibility as touch, taste, or smell.[45]

Above all else, the message in the mature stage of training nuclear workers was "believe the instrument," to suppress doubt – "This thing can't be reading right" – about the symbolic register of the dial. Stephen Frost, who began at Point Lepreau as a service maintainer straight from high school in 1982, doesn't remember any cowboys in his time: "Fear was probably at the root of it. I mean, I can't see this, I can't smell it, I need to know how to use this instrument. So I'm really going to pay attention here."[46] By the 1980s, the knowledge the pioneers had done without was available. The stations had more and better instruments, and workers' fear was amply informed.

Burnham's take on fear was bawdy. His students had experience with demanding manual work. The hazards from which he hoped to protect them could inflict grave bodily, including genetic, harm. He interspersed the conceptually difficult content with groaner word play and pictures of himself, "making a fashion statement," appendages inserted into the body scanners, face obscured by visor and hard hat, body clad in plastics

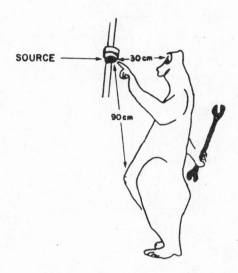

FIGURE 3.2   "Friendly Maintenance Supervisor," a cartoon that reminded nuclear workers of radiation exposure as a reproductive hazard | Jan Burnham

vacuuming up contamination (See Figure 3.5).[47] Physical humour brought the body back into a setting where bodily sensation was ineffectual as a defence. On tramp contaminants in the tubing: "Usually, beta-gamma sources have enough shielding to absorb all beta radiation. But if they do not, beware! ... At point contact, the dose rate would be greater than 1000 Gy/h/ This is not a misprint. ALWAYS USE TONGS WHEN HANDLING SMALL SOURCES. Never use tongues."[48] On the personal dosimeters (TLDs) each worker was required to wear in radiation fields: "Your TLD badge is worn in a prominent position on the front of your body between waist and neck – most people clip it on their breast pocket rather than to their nipples ... If you are working near a source such that the TLD would not measure the maximum dose to your whole body, move the TLD to that part of your torso that will receive the highest."[49] The image of the friendly service maintainer speaks to another concern.[50]

Gallows humour acknowledged the shared physical threat and channelled the bodily response it invoked. Burnham claims that he would "stick a cartoon in" on the blank page opposite chapter heads to "get guys to look at the book ... everybody else treats it deadly seriously and its bloody boring so I threw a few jokes in."[51] Maybe, but look at the work these cartoons are doing in a training course where the learning curve was steep,

FIGURE 3.3 "Wish You'd Called Me Earlier" | Jan Burnham

FIGURE 3.4 "Tombstone" | Jan Burnham

the hazards unfamiliarly insensible, their presence only indirectly knowable, their long-term and landscape effects uncertain and (when the effects were bodily) potentially deathly.[52]

Burnham's pedagogy nurtured working knowledge, in this case a theoretically informed practice composed of finely honed interactions between the automatic and the attentive, actions learned to be instantaneously disruptable by conscious intervention. "Although instruments are essential to detect and measure radiation fields," he counselled students to "assess the magnitude of a hazard" by setting the results from the instruments within the context of the history and current operating state of the reactor. This was in order to "get a feel for the hazard in advance."[53]

Crafting this synthetic, embodied working knowledge was complicated. Yes, the hazard was insensible except through instruments and calculations. But workers' protective clothing put more of the sensuous beyond ready register. Gloves limited touch, though surgical gloves were permissible for some jobs, and work around large equipment did not require fine sensuous discrimination. In areas where tritium levels were high, however, and tritium levels rose steadily through the life of a reactor, workers wore plastic suits. From their earliest versions, these limited vision and hearing. The first suits were unventilated, and, in the course of a job, a man would sweat several pounds of water, a load he carried about his lower extremities as he climbed ladders in and out of the reactor. The teams who wore this apparatus at the NRX, the NPD, and Douglas Point were all young men. Older workers could not have borne the physical assault. Over time the technology improved but with mixed sensory implications. Once the suits were ventilated, their wearers became dehydrated, a state that compromised both comfort and concentration. Blown out several centimetres from the body to offer the protection of positive pressure against ambient hazards, the suit literally gave men tunnel vision, and the rush of moving air, especially around the face, was disorienting. The roar of the ventilation system made hearing even more difficult, and the heavy shielding around the reactor meant that supplementary electronic communications were never very effective. The haptic constraints were considerable. The air hose was a tether that restricted the range of movement. The inflated suit was big and cumbersome, giving a man an unaccustomed girth and an ungainly gait, complicating access on the ladders and amid the tight spaces of the station's active zones.[54]

The suits were a new and foreign environment, a radical alteration in "spatiocorporeal field" for the wearer. Men were aware that they were not getting the "same kind of feedback" through their bodies while they were

FIGURE 3.5   Jan Burnham in plastics vacuuming | Jan Burnham

in plastics, that they "lost some of their senses" and "were not as well oriented." Journeys out in plastics required thorough planning, training, and practice. When a feel for the station's spaces, distances, and historical fields ordered so much of daily work, the bouts of near total insensibility in plastics, the couple of hours not knowing whether the atmosphere at the job was cold or hot, fresh or fetid, were discomfiting. Dan McCaskill, a mechanical maintainer employed at Point Lepreau from the beginning, concerned that he might thus have been a danger to himself or to others, compensated: "I think you're much more aware of what you're doing if you are doing it in a plastic suit. Because you take nothing for granted." Being deprived of many sensory reference points by the protective equipment focused workers' attention and made them more conscious of their vulnerability and mortality.[55]

By the 1970s, most workers, unless they were fuel handlers, spent relatively little time in plastics. On entering active zones, places where contamination might be present, they donned "browns" (the canvas coveralls the station provided), safety shoes, issue yellow socks and underwear, a hard hat, and safety glasses – garb not so different from that in other process industries. For some tasks, the ram's-horn-ventilated hood and jerkin worn over browns, which limited the inhalation of contamination but not its absorption through the skin, was a practical compromise between bodily sensation and protection. In browns alone, workers were no more hampered gathering sensory information about other features of their job site than were workers in other process industries.

From the NPD onward, there were four zones in increasing order of expected radiological hazard in CANDU stations. People who entered the plant wore radiation badges colour coded to the highest zone their training qualified them to enter, their physical mobility limited by their theoretical and practical training and signified by the colour of their badge and surroundings. At Point Lepreau, the zones ranged between likely safety (orange) and certain danger (red). Workers saw the colours through their radiation knowledge. Dan McCaskill recalls the colour of Zone 2, where the radiation danger was lower, as a "pretty blue," and of Zone 3, where most work in radiation fields was done, as a "hellish green." Tubing, too, was colour coded in order to distinguish pipes carrying heavy water that might be active from those carrying coolant water from the nearby Bay of Fundy.[56]

Engineering design, the location of the reactor, and the paths of the moderator and heat transport systems dictated the zoning of the station. A map of the zones was located by the entrance to the reactor building. Workers took as known and incorporated as reflex this topography, which was relatively stable in intensity and composition.[57]

Passing from zone to zone involved a good deal of body work: registering the levels on personal dosimeters by inserting them into a pad reader on the way in to change clothing, collecting and calibrating instruments, and assembling the materials needed for the job plan. On the way out, body work involved carefully disrobing, pulling off each garment inside out to contain contaminants, and washing down and bundling any equipment that had to be removed for repair so as not to leave a trail of contamination along the route to the shop. These routinized actions of approach and repair gave the radiation fields an embodied presence.

Upon this topography was played out a less stable and more finely grained history. Within the active zones 3 and 4, every completed job plan recorded the dose taken. These were the radiation history of the task, the

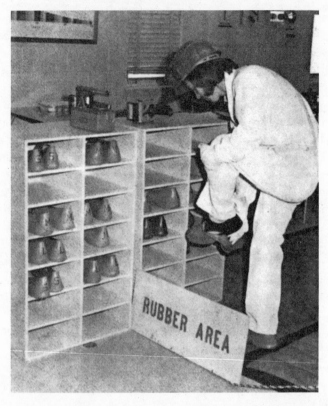

FIGURE 3.6   Changing shoes at boundary of potentially contaminated area | Jan Burnham

planning foundation for each person subsequently assigned the job. And the station had two sorts of history: the rising secular trend tritium levels in active zones and the cyclical changes in fields as power levels were varied for fuelling and outages. Knowledge of these histories informed workers' expectations as they moved about the station on a given day and gave them what Burnham called their "feel for the hazard."[58] How they would actually work that day depended upon the breaking news of system levels and "hot spots" conveyed by stationary monitors and field operators. This current information about the fields was posted outside the work control and radiation control offices, at the zone boundary and on chalk boards, by type, time, and intensity at the point source in the zone. After 1992, workers used it to set the visual and audible alarms on the gamma instruments they wore and carried into Zone 3. If the field they met was different from

FIGURE 3.7   Rubber station at boundary between zones | Jan Burnham

the field they had schooled themselves bodily to expect, sound and light jarred them to a different somatic mode of attention.[59]

Within zones, workers established their own boundaries as necessary to contain the contamination released as they opened equipment. They made a rubber area by laying down a canvas floor and erecting physical barriers and signage to direct other workers in the zone away from particulate contamination on surfaces near the worksite. If the hazard being released was ambient, the person assigned the job built a ventilated tent and worked within it until the system was resealed.

Light water and graphite stations in France, the United Kingdom, and the United States are also bounded internally, the transitions in insensible dose given visible, audible, and haptic markers. The response to these boundaries is a patterned sequence laid down in procedures and incorporated as habit memory. Studies completed in the 1980s in these jurisdictions suggest that the work these markers do may have been different when workers' knowledge of the dispersion, penetration, and decay of radiation was slight, when the markers and procedures were received as shallowly grounded demands for compliance.[60] A worker with a richly informed understanding of insensible radiation fields more likely accepted such signifiers as counsel for deliberative action, and attended to them and embodied them without a sense of personal violation. The different knowledge bases and authority structures have sustained both different somatic modes of attention and different risk behaviours.

Similar somatic implications follow from the early decision to minimize the use of contract workers in CANDU plants. The onerous days at Douglas Point convinced Robert Wilson, the health physicist in charge, that stations must be designed and managed so that they shed no more radiation each year into human flesh than the regular staff complement allowed by international limits. His goal was to eliminate dose fodder, the use of contract workers from away, and staff borrowed from other parts of the station to do jobs after all the regular operators and maintainers had exceeded their allotted doses, people who took dose and then slipped from place and mind. The effect of the fields would be borne in bodies of familiars, its accumulating impact apparent in the work roster of the station. A supervisor or first operator so unlucky or lacking in a plan as to "burn out," to exceed his dose for the period, would no longer be seen in active zones for the duration. The multiple competences Lorne McConnell required of each nuclear worker from the NPD helped here, but no tradesman, whatever his competences, could take over from another another unless he himself had current allowable exposure to expend. Some work, on consideration, could be deferred until the next dose period. Sometimes the only way to get a job done in the highest fields was to distribute the dose, sending in members of a shift crew one after another for fifty-second turns at the crank on a valve. Personal data boards in the station were public records of who had taken what dose. Managing contact with radiation fields in this way gave the bodily implications of the insensible hazard known limits and locations. This was not a burden imposed by stealth and treacherously distributed but a diet of dose, an ordinary individual responsibility with collegial implications.[61]

The working knowledge that guided operators and maintainers in Canadian nuclear stations was grounded in radiation theory, which described the topography and morphology of the insensible radiation hazard. The likely intensity of fields was sensuously displayed. The station was colour coded and barricaded so that passing into high fields of danger required physical performances, including the donning of prostheses that extended reach and protective clothing that reordered the sensing body. Shift routines honed habitual paces and paths of movement that embodied this longitudinal knowledge of the stations' fields, and workers learned to practise modes of somatic alertness specific to the visible and audible signs that current hazards might present. They spent decades year round ranging about in the same station registering its fields through maintenance and fuelling cycles. They knew the dose they took each day and the dose being parsed for current work among the staff complement. The historical and technological foundations of Canadian nuclear work culture made individual responsibility for radiation protection plausible and practical. Responsibility concentrated the attention that theory informed. Thus, the intimate reordering of somatic attention to accommodate radiation fields was about respect for self rather than submission to authority.

http://
megaprojects.uwo.ca/Iroquois

On this site, you will find a Flash-based presentation of the town of Iroquois, Ontario, as it was before its destruction in the 1950s. This is not a virtual reality-style simulacrum of the town but rather, by displaying several maps, expropriation photographs commissioned by Ontario Hydro of each house and building in the town, and audio recollections of former residents of walking through and living in the old town, a visual and auditory experience of the old town as remembered by those who were there.[1] The photos of the houses and buildings are arranged as they were located in the town, synchronized with relevant audio recollections and oriented to various maps. Thus, the user is able to "wander" along either side of both prominent Iroquois streets, King Street and College Street, as well as visit locations of interest to townspeople, such as the Caldwell Linen Mill and the waterfront areas.

Many of the former residents quoted in the pages that follow can be heard in the audio clips on the site, which, along with the pictured streetscapes, lists of property owners, and related photographs, contextualize and bring together the sometimes fragmented memories of the old village. The trees of old Iroquois, the "palpable evidence of persistence" that many townspeople weave into their accounts, are evident in the photographs on the site. Photographs of the mill, along with other mill recollections, are in the section of the site called "Explore Caldwell's." In "Other Old Iroquois Photographs," many of the particularities of Iroquois recollected in this chapter can be seen and heard, including arresting photographs from Iroquois, mid-destruction, with trees cleared, the massive house moving machine employed to relocate houses up to the new town of Iroquois, and a contemporary aerial photograph of the "model" new Iroquois.

—*JV*

4

MOVEMENT AND SOUND

A Walking Village Remade:
Iroquois and the St. Lawrence Seaway

The megaprojects of the postwar years were true to their name. The military bases, hydro dams, nuclear generating stations, and their allied transportation corridors and process industries were massive, and in their wake followed daunting and unfamiliar clashes of scale. Along the Seaway, neither vegetation nor landforms could persist against the huge machines moving earth to create hydro dams and deep-water channels to take ocean-going ships to the centre of the continent. Nor would the plans of the Canadian state be much altered to accommodate the preferences of the people who dwelled along the river. The residents of territory that became Base Gagetown cashed their cheques and made their own compromises in the unfamiliar circumstance of having liquidity but no land. Many who left agriculture and forestry chose to settle nearby in the villages of southeast New Brunswick or the city of Saint John, able to some degree to maintain their former ties through the lodges, churches, and women's institutes. But those looking for land had to relocate in other parts of the province upriver from Fredericton or in the northeast, away from the communities they had sustained and that had sustained them. The farms and urban centres where they began again were different from those they had made along the Lawfield Road but not entirely unfamiliar. Theirs was a deeply con-strained choice among built forms and landscapes they knew. Along the St. Lawrence, the challenges to senses of self in place were different.

Iroquois was the first of the eight Ontario villages along the St. Lawrence River remade to accommodate the St. Lawrence Seaway. And remade it was. Its stores and garages, streets, sewers, schools; places of worship, work, and play; the houses that could bear it and their residents, whose forbearance was not questioned – 1,049 people[2] and 151 structures[3] – moved 1.6 kilometres north from the old river's edge onto pasture land beyond reach of the coming flood. The faces of neighbours were recognizable, but beyond that, the Seaway project had removed the reference points through which the St. Lawrence villagers had made sense of themselves.

Cooke Sisty's Italian-born father and grandfather had helped dig the nineteenth-century canals past the Galop Rapids at Iroquois with pick and shovel, and they stayed on in the village as proprietors of a market garden.[4] In 1956, as heavy equipment operators made their way through the blasted fragments of Iroquois Point, which had extended south of the village into the St. Lawrence, Sisty guided his father, then eighty years old, to the viewing platform by the site.

> He walked on crutches and he got up those stairs very carefully and he got up on the top and I could see him look. He reached over and patted me on the shoulder, and I'll never forget him, and he said, "Son, they're changing the face of the earth."[5]

By the summer of 1958, the early nineteenth-century river village was gone, its remnants submerged beneath the rising waters of the pond behind the Robert H. Saunders-Robert Moses Dam. In its place, but 1.6 kilometre north, in a form transferred precisely from the drafting boards of the Hydro architects and planners, was a model town. There a modern plaza, its back to the village and fronting onto a parking lot, engaged the promised traffic of the new highway. Beyond, partially obscured from view along an incongruous suburban sequence of curved streets, villagers resettled in transported dwellings, shorn of their porches, perched on new foundations. New bungalows, austere in their promised convenience, bordered the planners' projection of a park.

In the 1930s, from his New York hospital bed the French sociologist Marcel Mauss observed nurses walking to and fro. As he happily admired the movements of the women who cared for him, he recognized a gait that Parisian women must have learned from actresses in American films, and he confessed a "fundamental mistake" that he had made and that he had gone "on making for several years" – "thinking that there is technique only when there is an instrument" when, in fact, "before instrumental techniques is the ensemble of techniques of the body." This active bodily product of "constant adaptation" to the constraints of social and geographical location is also technique.[6]

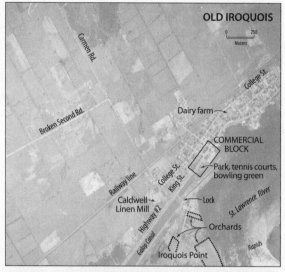

Sources: Her Majesty the Queen in Right of Canada, Department of Energy, Mines and Resources. Airphoto #A12449-122, 1950, Composite 2009 created from Google Earth (Image © 2009 DigitalGlobe, ©2007 Google).

FIGURE 4.1 Air photos of old and new Iroquois, showing the loss of the orchards, the rapids, and the propinquity of the water | Cartography Section, University of Western Ontario, Department of Geography

The observation platforms commonly built by megaprojects urged citizens to admire the technological sublime that machines were making in their midst. Yet the residents of Iroquois were caught unawares by the challenge posed by changes in the "face of the earth" produced by the Seaway. The bodily techniques they had honed unconsciously by daily interactions with the spaces, sounds, smells, light, and shadow of the old village misled them in the model town that the Seaway Authority had set down for them. Before the Seaway, Iroquois was a pedestrian place, "just a stroll-through town; people walking back and forth."[7] By walkers' measure, between the tracks of the main line of the Canadian National Railways to the north and the front, St. Lawrence River to the south, was a space twelve blocks long and two blocks wide. Although villagers frequently used boats to fish and for fun, and on occasion boarded trains for excursions to nearby Brockville, Cornwall, and the more distant metropolis of Montreal, for women and men, adults and children alike, walking was a "commonplace and ordinary" necessity rather than the "occasional, eccentric and symbolic" volitional activity it has since become: "When you wanted to go somewhere, you walked."[8]

A walker has little ken of "time-space" compression,[9] for walking is a mode of movement that "coagulates time, expands distance and makes places dense and prickly with details and complexities."[10] No matter how bundled against the winter or how weary after a long day at the Caldwell Linen Mill (the principal employer in the village), walkers were open bodily to the topography, the built structures, the vegetation, and the weather as well as to the other inhabitants of the village. Their sensuous and emotional histories in place unfolded unbidden as they passed along familiar streets, their "particular orientations toward the material and social world"[11] physical elaborations of their footfalls along the nineteen kilometres of sidewalk in the village.[12]

There was and is what Paul Adams has called "a peripatetic sense of place." To walk is to become involved by "sight, hearing, touch, smell, the kinetic sense ... even taste" through a "spatial practice" that opens up the human body moving and pausing, responsive to the setting.[13] "If you wish casual opportunities for meeting your neighbors, and for profiting by chance contacts with acquaintances and colleagues," Lewis Mumford wrote during the same years that the Seaway was being built and lamenting the rising hegemony of the US highway system, "a stroll at two miles an hour in a concentrated area, free from vehicles, will alone meet your need."[14] Mumford regretted that the US government's announced long-term commitment to fund highways for privately owned automobiles would limit the future choices American citizens might consciously make about how to live.

As was typical of postwar megaprojects, the pace of events in the building of the Seaway defied deliberation. In Iroquois the future arrived swiftly and thoroughly. Residents broached the streets of the new town in shocked reaction, "with the alert and hesitant step of strangers and immigrants,"[15] unwilling or unable to succumb all at once to the shock of the new. The mental maps and unforgotten histories that organized the spatial and temporal lives of village walkers persisted to define the ordinary for years after the ordinary of streets that had made them were drowned by the rising waters.

Their habits did not serve them in the new town. Muddled by the thicket of reflexes, once reliable and now discomfiting, they were at a loss to say why. Like the anniversary visitors who eagerly returned to their lands that had been remade as the military training grounds at Gagetown, they were unsettled to find themselves wandering about in the dark: "They missed their surroundings – walking uptown. Everything around them was familiar to them. And then we came back and there was a bog. It was quite a kick in the pants."[16]

A pedestrian pace is slow enough to capture detail. That's why the service maintainers and assistant operators in nuclear stations do their surveys (their term of art is "walks"), gathering information on foot. Walking is an exchange of information. Like the nurses in Mauss' New York hospital ward, walkers in public display themselves and, perhaps, betray themselves to scrutiny.[17] In the early 1950s, Jean Shaver was a young mother at home:

> Well, I don't think there were any places I didn't like to go, but I used to feel sorry for the girls that worked in the mill because they had to walk by our house and they would be going to work at six in the morning in the dark. And they'd walk right up the centre of the street and I used to feel sorry for them, because I used to think that was an awful time to be going to work to begin with ... they worked until three or something and then they'd all walk down again.

Joyce Fader, then a country girl boarding in the village and mastering her skills as a weaver, remembers these street scenes differently:[18]

> [Our recreation was] walking the summertime and just sitting around and chatting ... I remember another girl that I boarded with and I, we'd go to

work at six o'clock in the morning time and I can remember sitting out
under the tree – and talking all night, walking in the house, changing our
clothes and going to work ... we were always making plans, as girls do.

Keith Beaupre said, "There seemed to be more activity on the front street.
They had tennis courts in Iroquois and they had lawn bowling and people
used to go and watch them all night."[19] Before the Seaway, Joe Roberts, a
mechanic at the mill, "could walk to work ... the houses were quite close
together on the street where I lived and [it] was just a short walk to the
Caldwell Linen Mill." There were patterns to the place. Erma Stover, who
started working in the family store when she was eight, said, "I didn't have
a lot of spare time, but when I did, I knew exactly where to go to meet
some friends ... It was such a small town that you didn't need a telephone,
you just all showed up at the same place, like the ball fields at night or the
canal in the afternoon, or the rinks in winter."[20]

After, the village seemed hurried by urgent needs to reclaim the lost
familiar. Unlike dispersed farmers and lumberers of the Lawfield Road,
the residents of new Iroquois relocated as a group 1.6 kilometres north of
the old village. The human inhabitants were familiar, yet Roberts, unsettled by the radical changes in the built environment, no longer walked.
It seemed to him that "you didn't know anybody ... there was no togeth-
erness."[21] Or perhaps the connection worked the other way, for the
physical act of walking "alternately follows a path and has followers, creates
a mobile organicity in the environment ... like a series of 'hellos.'"[22] Physio-
logically, walking, "as the intentional act closest to the unwilled rhythms
of the body, to breathing and the beating of the heart ... strikes a delicate
balance between working and idling, being and doing," "produces nothing
but thoughts, experiences, arrivals." The displaced villagers were missing
the ordinary, magical mingling of "errand and epiphany,"[23] and their accustomed pedestrian practices bequeathed part of the disturbing absence
Ron Fader remembered, physically, as a "kick in the pants."

"The monotonous uniformities" Mumford bemoaned as the product
of "American zoning practice"[24] robbed the instant new Iroquois of the
ragged variety residents had savoured in the old village – dense complexities that had fascinated. The locks, the powerhouse, the park, and the
bowling green clustered by the old canal and the government ditch. The
large leather belts of the powerhouse were always buzzing, and Pat Beach,
the operator, welcomed curious boys.[25] Joe Roberts remembers being
introduced to static electricity at that powerhouse as sparks jumped three
or four inches (seven to ten centimetres) from the belts to his tingling

hand: "You'd think twice about doing it, but do it anyway, just to get a kick."[26] From houses by the water, that "could see the ships going up and down the river" by night, they could hear their engines throbbing.[27] Along the back street by the railway track,

> farmers in certain times would bring in their cows, pigs, sheep ... And there was a man by the name of Jim Lennox who ... had scales there where they were weighed ... And in the evening the trains would come in and ... they loaded the cattle. It was quite an industry at that time and it was something to watch too. There'd be a bull in there and [it] would get quite unruly and cause quite a disturbance. It was quite a thing to watch.[28]

In a shed behind a house on King Street, just east of the post office, Homer Everetts kept pregnant mares and from their urine cultivated pungent mould for serum. Nearby were a harness maker and a cobbler and a "place where they used to buy hides. You could smell them – hides up there – cow hides and sheep hides." And just west of the village there was a cheese factory where "you could reach into the vat and pull up a nice big chunk of juicy cheese,"[29] or have "a piece cut off the whole big round."[30]

Amid this fascination lurked a good deal of danger and inconvenience. King Street was also the busy main artery between Montreal and Toronto. The heavy traffic on the main CN line shook nearby houses: "I loved to hear that old train storming through in a way ... working twelve-hour shifts and trying to sleep in the daytime with that old engine snorting just opposite the house, it's a little different."[31] Caldwell's piped its process effluent directly into the government ditch, just south of King Street by the bowling green, colouring the water with dyes; the sewage of the village debouched into the river nearby, behind Bulles' garage, and dead farm animals, disposed of into the river by farmers upstream, were caught by the gates of the locks and had to be forked forward into the current. After coal for the linen mill was unloaded at the wharf, for several days trucks hauled loads day and night through the streets; the trailing dust blackened windows and seeped through the cracks into houses. For the duration, washing was impossible "because you couldn't hang anything out."[32] Seasonally, propinquity to the river brought shad-flies, darkening the street lights, their foul fishy-smelling remains accumulated centimetres deep by the morning.[33] Some of the boys who loved swimming in the canal by the powerhouse and the bowling green also drowned there, and near misses were common.[34]

The linen mill persisted after the Seaway, but on Iroquois Point, the three orchards of McIntosh apples (a species discovered and refined in

nearby Dundela), 2,200 trees yielding seventeen thousand bushels annually, the principal export of the village, were lost to the new canal. In spring
the fragrance and sight of the trees in blossom had delighted; in fall the
fruit hung bold red by the sparkling river. The harvest provided local
seasonal employment: "the older people picked the low branches; the
younger ones climbed on ladders and picked the top." Women graded
and polished the apples, wrapped them in blue tissue, and packed them
in the wooden boxes the men built for transport to the English market.[35]

Iroquois before the Seaway was a compact mingling of workplaces and
residences, a place of open porches and open doors. The technologies of
workplaces and households were not black-boxed, hidden from view;
rather, their processes, fragrances, stenches, and wastes were involuntarily
perceptible, and, perhaps in the case of the power plant and the lock, in
retrospect, voluntarily too accessible. Walkers, unarmoured by the bodies
of cars, habituated themselves to these complexities through the thicket
of bodily technique. But habit and reflex do not readily rise to speech, and
when residents strove to account for the dull anomie of the new village,
they invoked a new consumer technology – the television: "They stayed
inside; they had their programs to watch and you didn't want to interrupt
them by visiting."[36] The model townscape was so strange and estranging
that the passive pleasures of this new black box, which captivated so many
in the mid-1950s, may have been a particular solace and distraction in
Iroquois.

The villagers did not get the new village they wanted. But not for lack
of initiative and forward thinking by their reeve and council. In August
1952, only months after Canada announced that the Seaway would proceed,
and before the US Congress had agreed to the project, the village council
hired Wells Coates, an eminent British planner and architect[37] whose own
roots lay in nearby Prescott, to design new Iroquois. Coates chose a
comely place northeast of the existing village in Matilda township at Flagg
Creek, which he thought suited for the development of a commercial
harbour. The location had ample flat land for industry and heights from
which residents could have commanding views over the St. Lawrence.
Coates planned for a town of twelve thousand residents and set about
marketing the project to British manufacturing firms.[38] When the neighbours in Matilda refused to cede their farm land, Hydro engineers questioned the stability of the marine clay at the creek mouth, and British
industrialists were not forthcoming,[39] the utility assumed control.

The plan Hydro presented to the villagers in August 1954 was minimalist. "Hydro's new Iroquois could be set down in the middle of the prairies,"

Coates objected. Jack Fetterly, a councillor, perhaps showing the fatigue of too many eight-hundred-kilometre return trips by car along the two-lane roads to represent the village at Hydro headquarters in Toronto, described himself and his colleagues as "crest-fallen and disappointed given what they had been led to believe." The proposals Hydro presented were both unimaginative and ungenerous, and "the planners had chosen to ignore the important part the riverfront has played in the lives of the area."[40] The evening the models were unveiled to the village, citizens protested "they had been 'on the river' since the town was started and they should be there again."[41] But the revisions Hydro offered were meagre. The new village would be on the fields to the north, the old townsite filled with material dredged from the river bottom. Three-quarters of a kilometre back, the villagers were "going to be given a shallow waterfront," which they could only hope "could be controlled and kept clean." Hydro would not assume the expense of replacing the lost harbour. Faced with an offer they could neither countenance nor refuse, villagers were left on their own "to get the best possible under the circumstances,"[42] circumstances Coates, now back in London, England, described as "pedestrian in approach and suburbanite in concept," conceived in isolation "from any contact with the lands or the people."[43]

Managing without the river was going to be tough for everyone.[44] Their bodies were familiars of the river, their senses of self dependent upon time-worn knowledge of the St. Lawrence and its ways.[45] The river was "your scenery ... you'd go back to the back streets and the side streets, but you'd always be looking at the river over there, by these houses. The river was always there." Even for adults, pressed by responsibilities as parents and wage-earners, church and community leaders, the river was a vicarious pleasure, an enveloping, prospective solace: "I loved that water ... loved to see other people out enjoying themselves ... when I did find time, I just loved the river ... just the fact that I could go out and wander around and stand around."[46] Aurally, the river was the keynote of the village:[47] "Normally the river was fairly quiet, but you knew it was there." It had an ambient quality of shifting precedence: "I know at night many times my father's been fishing across the river and I'd go down to meet him, and ... I'd sit there – you could hear the river, it was quite rough and you could, hear it very plain." This speaker is Don Thompson, who like his father Samuel (nicknamed Tacker), the town's fuel dealer, lived and worked by the riverbank and built splendid boats. Sometimes the river was a signal: "You could hear it slapping up against the banks, rough days when it was windy." But most days the river was the barely perceptible yet comforting ordinary of the village soundscape, a kind of muted roar. This keynote had a deep and pervasive influence on the behaviour and moods of those who lived with it, always in the background until the day when the coffer dam, built to hold back the river so that construction could begin, was closed. That day the river and its rapids became thunderously inaudible.[48]

The current by old Iroquois was fast, thirteen to sixteen kilometres an hour at the village, sixteen or more in the rapids between the point and the American side. "It was an interesting river," to the lock-master's daughter, "and it was a bit treacherous too. You had to be careful on the river." If to the local walker the sound and sight of the river was familiar ambient, to the local boater and swimmer it was an embodied place of active affirmation: "A river like that, you've got to know it." "You learned to respect it." You needed someone to "show you how to row and how to hit the current properly and if you want to get up around a swift point when you're rowing ... it was a quick trick, to pick out the little eddies ... in a skiff, two pointed, at both ends." But once you had taken such an oath of solemn respect with the river, its waters were no longer forbidding natural elements. You shared in the river's power. The river became "interesting."[49]

The river drew upon and made valuable the deep reserves of craft skills among the people of Iroquois. As for the physician, the athlete, and the dancer, so too for the craftsperson: skill is a tuning of the physically stored processes of perception. Many men built their own boats, which, until the late 1940s, were of mahogany and brass screws; thereafter, they were often of plywood. From November, the lofts of many outbuildings along the back streets were turned into boat works, and, come spring, in public launchings, the products of this consuming and somewhat covert winter activity were revealed to the village. Friends who built boats together also fished together, mapping their accumulating knowledge of the fisher's river on wall charts in the sheds where they built boats.

When villagers identified themselves as river people, they were thinking of these aspects of their deep generational sensuous histories, those bodily habits nameable as skills. The boat-builders' skills could and did change with technology and fashion. By the late 1940s, some villagers had even turned to buying their boats. But the fisher's knowledge was specific. It pertained to particular points, to pockets in certain bays sheltered from seasonally specific eddies.[50] Who knew, anticipating the change, whether this knowledge would weather the coming transformation of the river?

By study, habit, and accumulating acquaintance the villagers had embodied the river, taken it unto themselves. Were we to follow Bruno

Latour,[51] we might say that they had moved the bracket between culture and nature a little, so as to claim more of the river as cultural kin. Mostly this was manly knowledge, passed from fathers to sons. This may partly explain why, when it became apparent in the summer of 1958 that the old river was gone, among the most traumatized were the older and the most water-wise men of the village. The event that transformed the river at Iroquois was abrupt and cataclysmic, but the sensuous history of the change was different. Some succumbed to the trauma right away. For others, in the summer of 1958 there began something akin to a long, testing, and unwelcome regimen of physical rehabilitation, a struggle back, which could accommodate the effects of the trauma but never quite restore to the self the sensuous body it once had been. Villagers who had shared the river's power now shared the river's loss.

After the summer of 1958, the river no longer passed near the streets and houses of Iroquois. For a time the water's edge was a wasteland, resistant to all living things. It has since been turned into a golf course and an air strip, a park in name but, in reality, an open windy space that yields few park-like pleasures.

When the villagers use the word "river" to refer to the water there at the end of the park, their intonations suspend the inflected sound between quotation marks. The maps name the body of water downstream from the remains of the point Lake St. Lawrence, and the villagers, particularly the families who had fished, refer to it as "that lake" or "that pond," as in "it's not the river anymore, it's just a lake – it's a pond." The new dams served their intended purposes. The rapids were drowned, the current subdued. The expanse of water was wider and, but for the shipping lanes, less deep. The richly knowable old river was gone, replaced by an abridged alternative that was "just a navigational body of water for shipping." The shallow waters of the new river-pond-lake blew up more quickly with the wind than did the old river and sooner became rough enough to be dangerous for small craft. The old river had been called beautiful and interesting, a "place that made you feel like a millionaire." The new river-pond-lake was unfriendly, its power bound, "all dammed up so that there was just nothing there." And the fish seemed to be gone. All the places where they were known to be were gone and, truth be told, the new marine habitat was as inhospitable to fish as, at the beginning, the new park was to people.[52] Many villagers no longer fished or took to the river with any enthusiasm. No persuasion from his son Don would get Tacker Thompson, he who had spent eighty years on the river, into a boat after the river-pond-lake came. The old river had taken the water-wise

FIGURE 4.2   Cooke Sisty's father with vegetable wagon on the north side of King
Street in old Iroquois | Sisty family photo

men of the village into its confidence and they had extended it the respectful familiarity due a good neighbour. After the flooding, that neighbourly presence was unreachable: "It's as if you got moved to a different
country – the river just never appealed to us in the same way."[53]

There were many places nearby where the old highway, still following
its old course, ran directly into the water. The sight remains unnerving to
this day. Until he died, not long after the inundation, villagers kept a keen
eye out for a respected elder, the former public school principal, who, at
a loss for familiar markers, sometimes wandered to and into the water.

Six months of the year, the villagers of Iroquois had lived under a sheltering canopy of trees so dense that it is difficult, from air photos, to make
out most structures. Many of the village maples were nearly a metre in
diameter. Their branches touched as they lined and arched over the streets,
providing physical shelter, shade in summer, a wind break in winter, and
year round a helpful menace that slowed Toronto-Montreal through traffic along the curves of King Street. The giant lines of ancient pines that
stood between the orchards and the river on the Point probably predated
the arrival of Europeans. Many of the maples were as old as the village,
comforting signs of continuity, continuing sources of pleasure, palpable
evidence of persistence.[54] Accounts differ about when the trees came down.

Some remember the village being clear-cut, houses transformed by the absence of the canopy, gardens baking in the sun, months of chainsaws running, slash burning, and people weeping as the maples fell.[55] The dated photographs from the expropriation phase suggest the process was more gradual. Fresh stumps, ten centimetres high and nearly a metre wide, are visible before the sharp contours of inhabited dwellings in foregrounds, while in the blocks behind, the softening cloak of leaf and branch remained. In the end every tree would have been felled because come the flood, like the buildings, they posed a hazard to navigation.[56]

In each of the case studies in this book concerning the human neighbours of megaprojects, the issue of compensation looms large and complex as much of what would be taken away to make way was not fungible, and had never been intended for sale. This was not entirely the case for the woodlots at Gagetown. There the issue, at least partly, was timing, the surrender of the legacy timberlands before they had reached the merchantable maturity for which they were intended. In Iroquois, trees and people were both expected to die where they had lived and then, with due solemnity, be carried away and replaced by the next arboreal and human generation. Les Cruickshank, in the 1950s a young contractor making good money on the Seaway project, remembered an older man from Iroquois saying:

> "I don't mind moving and so and so and so on," he said, "But what price do you put on trees?" And I think the guy had ... you know, an argument there. It was the price he paid for the Seaway. They picked up his house and moved it back into a field. And he said, "I'll never see trees again." And he never did. He died before his trees got to be a nice size.[57]

"It was like elderly ladies being stripped of their clothing in public, when the trees came down that had sheltered these homes for so long ... You hardly recognized them as being the same building."[58] Ron Fader's grandfather became so disoriented by the absence of the maples that he could not recognize his transported dwelling as home, and when his kin could no longer manage his confusion, he spent his last days in the psychiatric hospital in nearby Brockville.[59]

The new village was barren and windy. Without the canopy, people felt more distant from their neighbours and exposed to the wind and dust storms. They were anxious to protect their children from too much sun: "You know redheads and blisters."[60] And so: "Well, we went crazy with trees," "everything that would grow fast, Lombardy poplars, willows, silver maple ... they were all just like weeds." Unlike the great maples, which had been a sign of continuity, these "poor quality trees" were woody similes for the unbidden predicament of the village. They "outgrew themselves in twenty years" and had to be removed. Shirley Fisher took down twenty-nine from her lot alone. It was really fifty years before the village "began to look like something," that is, before it was treed with real trees, again.[61]

Since time immemorial the twenty-eight-metre fall in elevation over the one hundred twenty-two kilometres between Prescott and Cornwall had been an impediment to all movement; at many places along this stretch of the river, the entire flow out of the Great Lakes was compressed into narrow passages bounded by resistant pre-Cambrian outcrops, sending spray several metres into the air. The rapids along this distance were spectacular. Three metres of the drop occurred at the Galop Rapids just east of Iroquois. The gap just downriver south of Iroquois Point was only 500 metres. First Nations peoples and fur traders in succession had paused on the Point to strategize before proceeding. The setting also persuaded most early industrialists to seek locations further upriver or downriver as the natural turbulence of the International Rapids rang out as a disincentive rather than as an invitation to economic development.[62] The Seaway planners promised to undo this history by re-engineering the topography.

Their promises were grand, even grandiose. The Ontario minister of planning and development foretold that the area they renamed "Seaway Valley," astride the main highway and marine and rail lines connecting Toronto and Montreal, with thousands of acres of surplus land, was, "now an area of greater industrial potential than Germany's Ruhr Valley."[63] Two local councillors remember being "seduced by the promise of another Ruhr"; one held onto that dream for a quarter century.[64] But all along, indeed during the four hundred years Europeans had been trying to make a passable waterway through to the Great Lakes, there had been sceptics. Sir Adam Beck, the founder of Ontario Hydro, doubted the prospective gains of a seaway would adequately counterbalance disruption to long established settlements. Norman Wilson, charged during the Second World War to report to the federal Advisory Committee on Reconstruction on the development of the International Rapids for power and navigation, concluded that, "short of a controlled plan to decentralize industry in Canada," there were no reasonable grounds to expect that, after a seaway, Iroquois or its neighbours "would become more industrialized than [they] ... had been in the prewar years." Theo Hills, an eminent McGill University economic geographer, correctly reasoned that given the long winter freeze-up and the imbalance between Sept Isles iron ore cargos going

FIGURE 4.3 Sign by Daffodil Cafe, south side of King Street, the old lock behind. The sign to the right of the cafe reads: "WE HAVE TO GO ~ BUT WATCH US GROW." | Ontario Power Generation

upriver and grain going downriver, especially since these loads would not use the same carriers, the Seaway would never recover the massive capital expenditures its construction entailed.[65]

After their own plan, the Coates plan, was discredited, many in the village recognized, as did Ann Thompson, that Iroquois "would not get very much larger."[66] For them, the sign posted on King Street leading into the old village "WE HAVE TO GO BUT ~ BUT WATCH US GROW" was more a concession to Hydro's dominance than an affirmation of Hydro's claims, akin to "whistling in the dark" against a looming uncertainty. From the time the utility resumed authority over planning, Janet Davis, daughter of the diligent and tireless reeve, remembers being told at home that the new Iroquois would be "a retirement village ... it wasn't going to be a real business area, we went modern, but the growing part wasn't going to happen."[67]

"Go modern" they did, their new village incorporating "the most progressive features" of contemporary town planning, a microcosm of Seaway promoters' goal to "control a great river more completely than had ever been done before in history."[68] The ragged variety of old Iroquois was succeeded by "coherence" and "consistency" in the new.[69] The village was planned. Wits said then it looked like a plan; in some ways it still does. New Iroquois was "neat and modern," perhaps "a little too neat." There was "an unnatural homogeneity" about the place that a visitor in 1969 worried "tended to increase the sense of the depression in the area."[70] The program called for mass housing, "economical, functional and attractive – in line with present-day trends," offering "privacy and quiet" rather than the variety and activity so savoured in old Iroquois.[71]

The locals were unimpressed: "We don't want to live in small standardized houses," they said. "We want houses as big and individual as the ones we have now." The architects' drawings were dismissed as "chicken-coops," "Southern-type homes, with open garages that would be of little use in this climate." Most were single-storey bungalows, at odds with the local custom of going upstairs to sleep.[72] As important, residents were not confident they would "be able to afford to live in the well-planned model communities" Hydro had "so glibly" proposed, given that the expropriation offers for homes in the old village fell well short of the costs of replacing them with any new dwellings.[73] Most villagers, living on the wages of non-union textile workers, were not prepared to assume mortgages on modern houses that did not appeal to them, and in a new village they were uncertain would ever be home.

And so, in a compromise driven by costs, in a climate of growing mutual antipathy, Hydro got the site and the site plan it wanted, and most villagers kept their homes, which were moved north using a house-moving technology pioneered in the American South for an earlier megaproject financed by the Tennessee Valley Authority. Wealthier residents built new, but to plans of their own choosing, mostly along Elizabeth Street, facing the park. Working people tolerated losing their porches, lean-tos, and outbuildings in trade for new water, waste, heating, and electrical systems and high basements. Some even allowed Hydro to paint their displaced dwellings in "brave apple greens and other bright colours" rather than the "off-whites and creams which [had] marked the community for years."[74]

The logic of the Seaway was the logic of a *mega*project. Modernism was framed for the metropolis. Scaling down is not an easy challenge in any technological process. Hydro's site plans for all the settlements in the Seaway Valley addressed city concerns and ignored village priorities. The result left residents contending with "a kind of spatial limbo," a suburban subdivision, with looped streets to reduce traffic hazards (in the absence of traffic); with curves rather than corners (in a community of exuberant gardeners who savoured the horticultural advantages of the corner lots common in "the old grid of small blocks"); with a plan designed physically to maximize household privacy (in a constituency that, having learned from childhood to mind their manners, avert their eyes, and keep their peace when in public, found the planners' suburban artifices an impediment to their traditional pleasures in shared space).[75] Because the utility, in the manner of contemporary megaprojects, emphasized control, eschewed local consultation, and planned in camera, Peter Stokes, the renowned restoration architect who did much to salvage the Loyalist-built heritage of the area, discerned with admirable understatement that Hydro's "improvements [went] largely unacknowledged" in Iroquois and the other new towns born of the Seaway.[76]

Even though technologies can resist changes in scale, they are often defenceless when users' resolute purposes differ from designers' intentions. Cultural geographer Yi-fu Tuan, writing about the moving traffic in the commercial districts of Los Angeles, describes the anomie of "the modern thoroughfare [where] there is no contact, for each person (or each small group of persons) is encased in a motorized metal box,"[77] intent on purposive expeditions to consume. As Rebecca Solnit notes, "Shopping centers proliferated in the United States, Canada, and Australia after World War II as wealthy populations of car-based families situated themselves on the outskirts of towns."[78] The form set down in the Seaway villages was novel in the mid-1950s, "a miniature of the giants in Montreal and Toronto,"[79] and fundamentally unsuited to the ways the people of Iroquois used both their commercial space and their cars. Modernists believed that form should follow function, and when they ignored their own postulates, as they did in the Seaway villages, they perpetrated an ill-chosen form that eroded healthy and happy socially generative practices. In Iroquois, this erosion still occurs. Jack Fetterly, who besides being a councillor in the village ran a business, observed the transition close up. In old Iroquois, "the big night was Saturday night when the farmers came to town, the stores stayed open until ten o'clock or even later, and the people would come in and visit. You would have four or five people in visiting. It wasn't necessarily much business, but they would be congregating and talking about things that happened." Villagers joined in. "It's just everybody on Saturday night, it was quite a thing to walk up one side and down the other, you'd stop visiting."[80] Sheltered by the three-storey buildings that lined King Street, under the trees, along the sidewalks, within sight and hearing of the river, shopping and visiting were mingled. And cars played their part: "All the people from the farms ... would all come in and they'd all park along the road and go shopping and then visit." Here, too, villagers joined in. Les Cruickshank remembered his father-in-law, proud of the brand new Dodge he bought every year, each Saturday going down early to get a good spot, coming home for supper and then going back

FIGURE 4.4　The commercial section of old Iroquois, looking east along King Street, Highway 2, a place to meet and linger | Ontario Power Generation

with his wife to sit in the car and visit. In this practice, vehicles were not speeding closed metal boxes. Stationary, in a space of vibrant social activity, with the windows and doors open and people clambering in to sit for a time and then move on, they were more like salons.[81]

Certainly, with the plaza the economics of village retailing changed. Thus Jack Fetterly observed, "The cost of doing business was considerably more ... so that you had to be more energetic and maintain a profitable year end." "The way the new stores were set up, there wasn't the extra space there and the stores owners had to use it" for merchandise rather than for the "pot-bellied stove and a circle of chairs,"[82] which had accommodated and encouraged lingering in the retail spaces along King Street.

The form of the plaza, with its arid parking lot facing the open frontage of the new highway and its back to the village, amplified these effects. People "came in and bought their groceries and they left because how were they going to see people? It just wasn't convenient." There was "no place to congregate in the shopping centre"; "not much place to socialize." "You went from business to business and you did your business and you were finished." The commercial space "looked small; it looked efficient"; it was a "here's your hat, what's your hurry" sort of place.[83] Whereas cars sitting along King Street had been sheltered overhead by a canopy of trees and

FIGURE 4.5 The plaza and parking lot in new Iroquois, south of the relocated High-way 2, also looking east, a place from which to hurry home | Ontario Power Generation

on two sides by the elaborate facades of Victorian buildings, and had interrupted views of the St. Lawrence, the plaza parking lot was exposed to the prevailing northwest winds, with uninterrupted views of the empty space allocated for "future development" on the far side of the highway. The parking lot provided few places to park, at least in the sense that the habits and bodily techniques of villagers practised parking: "If you could park in front of the stores, there would be a little bit of that left yet; but [not] if you're over there in the third or fifth or tenth row." "People come to town, buy their groceries and go right home."[84] Villagers' adaptations to the new built form are remembered negatively as losses rather than positively as efficiencies: "'You're back so soon. How come you're back so soon?' 'Well, I didn't see anyone.'"[85]

The nearest house to the plaza was three hundred metres away; the village had only one part-time constable. Thus, the new commercial space, because it was so uninviting to villagers, became an opportunity for strangers. There were five robberies in the first three months after the plaza opened; in the most daring, thieves smashed a store front and, using a tow truck, hoisted out a two-ton safe containing the weeks' receipts. Jane Jacobs would not have been surprised by this implication of absent neighbourly surveillance.[86] Neither was Les Cruickshank: "Looking back now, I can see that architecturally we could have done a lot better ... you know, give them a little bit of the old main street, there was something there to be retained ... we just hit the wrong year for that, maybe, architecturally."[87] Perhaps, but the minimalism of megaproject architecture was related everywhere to the postwar drive to expand national incomes and, in 1950s Canada, the United Kingdom, and Scandinavia, the goal to redistribute these income gains more equitably to the citizenry through welfare state policies. The people of Iroquois lost out to this fiscal triage, first when the Coates plan failed and again when national and local priorities collided at the place where their new village met the new river.

Residents wanted "nice, fresh running water, down right in front of [them]" along the new front street, a waterfront that would be some simulacrum of what they had lost, a waterfront they knew how to use and would use. Instead, the revised Hydro plan provided a 1,000–1,600-foot-wide band of open space between the St. Lawrence and the village, 200 acres (eighty-one hectares) in all. "Is this what the people want or what they are going to have?" an irritated villager asked in July 1956. Come October the reeve could only report that, "whether we like it or not, we're going to have a park."[88] Villagers were going to have a park because Hydro had accumulated a mammoth surplus of greasy glacial till blasted out of Iroquois Point and resistant marine clay dredged up from the river bottom. The cheapest option was to dispose of it nearby, and the front at Iroquois was nearby. The debris in some places was so dense that, in hours, it wore out steel blades tempered to last for years; elsewhere it presented as "a silly-putty nightmare," which when wet defied traction except when covered with cinders hauled in from Toronto and when dry emitted fine particles that caused "equipment to cough, sputter and shudder to a halt."[89] Into this unprepossessing humus-free base, Hydro planted myriad non-native trees: Serbian Spruce, Austrian pine, English Oak, Norway Maple, Spruce, Ginko, Beam, European Beech and Ash, and Linden, and numbers of native birch, willow, and maple trees. Arboreal migrants and natives alike were as challenged by their designated habitat as the humans struggling to resettle just north of the park. Those residents along the front street, Elizabeth Drive, who had been concerned the plantings would obscure their views of the river, need not have worried.[90]

The humans of Iroquois adapted their bodily rhythms and practices to the stage setting and choreography of the new site, as humans do. With some sadness and irritation, catching themselves betimes using reflexes that better fit the old than the new village, they absorbed the hurried-up sense of time. Joe Roberts drove rather than walked to work at the Caldwell Mill. Rose Sisty, Shirley Kirby-Carnegie, and Joyce Fader shopped efficiently and quickly returned home. People watched their new televisions and scheduled their weeks around programs they especially favoured. Some even took up golf for a while. Many regularly began to play cards at the former Catholic church, repurposed as the Legion hall.

But replacing the "second nature" they and their ancestors had crafted in the old village would not be hurried up: "I took a walk through our garden when we heard the news, and looked over all our shrubs that are in full bloom just now ... At the time they were planted with hope that we would see them in their full beauty and now, what happens ... to the plantings of orchards and small fruits, asparagus beds and foundation plants that take years to develop and produce?"[91] Hydro policy was not "to get into the topsoiling business."[92] The material around the houses in new Iroquois was the clay excavated for their basements: "It was a hard job to get it into shape, it took a few years, bringing in new soil and working it in." A good gardener needs to hold a long view of the passage of time and her place in the natural or "second" natural world.[93] In new Iroquois such a patient stance of stewardship and legacy seemed naive. The people of Iroquois were older than most residents of the province.[94] Few villagers gardened with the same zest as they once had done.[95] The topsoil issue became a source of mordant humour in the local press when the *Sudbury Star*, the newspaper of the northern Ontario nickel-mining city that had grown on rock outcrops where gardeners were also beset by smelter sulphur emissions, launched a campaign to truck north to the Shield the "millions and millions of yards of black earth" about to go to waste under the rising waters.[96] Along the Seaway Valley, people took their jokes where they could find them – grounds for laughter, unlike glacial till, were in short supply.

Much of what made the old and new villages function well came north with the villagers. In this sense, the people of Iroquois were better off than those who were displaced by Base Gagetown and the reservoir that replaced the Arrow Lakes. A team of University of Toronto social workers who surveyed the project area in 1957, recognizing that "never before has an established community of this sort experienced so abruptly or intensely the sweeping impact of sudden change," described Iroquois after the move in the most favourable terms a social scientist of the 1950s could summon. They saw it as "a settled and homogeneous small community with relatively stable families, warm family life, reasonably close personal associations and good neighbourhood relationships, adequate community organization except, perhaps, for recreational purposes, no apparent deep divisions, and few pretensions."[97] Unlike nearby Cornwall, burdened by the stresses of transient megaproject boom times, where local labour, housing, and marriage markets were disordered by the arrival of thousands of young male construction workers, Ontario Hydro seemed to have done well by Iroquois: "In its physical aspects the move itself was evidently well planned and executed. Nor does it appear that the citizens generally suffered in the value of exchanged property. On the surface, in fact, the people of Iroquois seem to have taken the move in their stride." Puzzled that "the destruction of old Iroquois was not seen by ... [residents] as their special contribution to the great undertaking that is the St. Lawrence Seaway," the social surveyors, while themselves captivated by Hydro's promise of megaprojects, discerned "that, whether Hydro was at fault or not, the people of Iroquois had no real sense of being partners in the adventure of creating a new townsite." Since then, "If megaprojects, then megaprojects with due process, local consultation, and consent," has become the mantra. Then, the Toronto social workers were uncertain whether the "active participation of the residents would have produced a different and better result."[98]

They were spot on. Certainly, at the time, Hydro's imperious ways rankled. Yet, in retrospect, it is not the lack of consultation but the loss, not the disrespect but the destruction, that endure disruptively in reflex and memory: "I must say that I think Hydro lived up to their promise very well." This is the boat builder, once also a fisher, Carl Van Camp. "But being wrenched out – I don't think people will ever get out of that." When asked, "Wrenched out?" he replied, "Wrenched out of their method of living."[99]

There was no going back. The people who were displaced when old Iroquois was drowned by the rising waters of the Seaway, like those from the Lawfield Road whose woods and meadows were thoroughly remade to serve the needs of the military, had no place to go: "Your home is just gone and you can't go back to where your home was, because the land is not there. Most people can go back because the land is still there, whether the house is gone or not, but the land is there. But when the land is gone, then that's a different story ... you can't go back, so you just have to keep going ahead, regardless of what happens in your mind."[100]

The villagers had made sense of themselves through their sensing bodies. After the flood, the fires, and the clearing, there were no physical reference points for the selves they had been, no benchmarks for the spatial practices of daily life, for the habits through which residents had embodied the place. Externally, "all of a sudden, everything starts working differently." The spaces in which their bodies moved changed. What had been near and there was now farther away and in a different direction. Light, shade, and shadow altered as most houses in the new village faced east-west rather than north-south (as they had in old Iroquois). The trees, the river, and the gardens were gone. Internally, the disorder was a deafening clamour: "I guess the quietness, that's the only way I can explain it. It seemed to be so much quieter in the older town. I don't know, maybe it's just all in your emotions, your mind, that you were quieter." With links between being and doing severed, between acting and knowing disarrayed, "there just never seemed to be that easiness in your mind ... It's just a feeling that you have inside you. You have no control." Politically, this loss of control was amplified by the regulatory processes of their displacement, the facades of consultation and the dissemblance of consent common to the military training base at Gagetown, the Seaway, and the Columbia River projects. But sensuously, perceptually, cognitively, the loss of control manifested as the depression so many have observed in the neighbours of so many dam sites. The people being relocated had grown and thrived as sensing beings by actively living amid the sights and sounds, pleasures and inconveniences, of places then destroyed in the name of progress. Consciously, learning that the Seaway "was obsolete when it was done" nettled: "You wonder why it was ever started if it was going to be obsolete when it was finished." Unconsciously, the awareness that was once so serviceable but that now registered against an absence was more difficult to place:

> Like, say, you have a car accident and you're fine that day, and then the next day, you've been in an accident and everything is different in your life, so completely different ... You go down in your mind. Where do you hold it? Where do you keep it?

A burden beyond speech lingers, testing resilience, haunting like a ghost.

http://
megaprojects.uwo.ca/ArrowLakes

This site presents the new media work associated with the damming of the Columbia at the Arrow Lakes. There you will find two main groups of material: first, a collection of photographs, audio, and video associated with the affected valley lands surrounding the Arrow Lakes, with particular focus on the Spicer market garden near Nakusp; and second, a chronological archive from another former Arrow Lakes farmer, Val Morton, documenting in his own words the interrelations between his life and ranch.

The story of the Spicer family and their land is enriched by the beautiful photographs shared with us by the late Jean Spicer of their once gently sloping waterfront land and family homes. We have added photos of Spicer Rose Gardens and Janet Spicer's fields at Twin Lakes and East Arrow Park. Using GIS data and photographs taken since the flooding, we have created imagery depicting how the post-dam waters submerged the former farm lands. There are also video clips available of Mike Halleran's CBC documentary on the Columbia treaty, *The Reckoning*, which includes video of the landscape changes along the Spicer's shoreline and depicts the fiery fate of the SS *Minto*.

The site also contains a new media project about Val Morton, a former rancher based on the Arrow Lakes several kilometres south of Nakusp and the Spicer farm. The Val Morton project was intended, like the Iroquois new media project, as a Flash-based *re-collection* of various recordings and materials. This work focuses on the making, over a lifetime, of the Morton ranch, and is organized according to succeeding ranch epochs from Morton's own reckoning.[1] Using his voiced and handwritten accounts of the year-to-year, day-to-day activities on the ranch from the 1920s to his expropriation in 1970, the site shows how Val Morton characterizes his own life on the ranch at a later time when, living in a camper, surrounded by photographs, newspaper clippings, and legal documents concerning the expropriations, he mournes its absence.

—JV

TIME AND SCALE

A River Becomes a Reservoir: The Arrow Lakes and the Damming of the Columbia

The things you had no right to do, the things you should have done,
They're all put down; it's up to you to pay for every one.
So eat, drink and be merry, have a good time while you will,
But God help you when the time comes, as you
Foot the bill.

These are the last lines of "The Reckoning," a poem written by Robert Service, a bank employee in Dawson City during the Yukon gold rush. Banker though he was, perhaps because the landscape of his dwelling place was being turned upside down and inside out by a mining boom, Service reckons needs and values no cashier's ledger would accommodate and sorts credits from debits on grounds no teller would recognize. He ties actions and obligations more fundamentally to moral rights than to statutory rights, and he considers accounts and accountability in social, sensual, and environmental as well as market terms. His is the classical economy of Adam Smith, a world where "value in use" – utility – the satisfaction of needs and desires, exists before and apart from exchange. This reckoning is of a large and heterogeneous world of goods, greater qualitatively and quantitatively than the sum of products and services included in the national income accounts.[2] Teachers often introduce the modern history of the British Columbia Arrow Lakes and the 1962 Columbia Treaty with the United States by showing a film that takes its name from this poem.

The Reckoning is the work of Mike Halleran, whose professional life as a filmmaker for the Canadian Broadcasting Corporation was undone by his determined opposition to the Columbia Treaty. The outstanding reckoning Halleran invokes is the loss of scarce, limited, and useful (although unpriced) fish, fauna, and human habitat as the waters of the Canadian Columbia River were turned into storage reservoirs for the power dams south of the forty-ninth parallel. The images of farm homes and buildings being torched, and the SS *Minto,* the sternwheeler that served the valley inhabitants, sinking in a flaming Viking funeral beneath the surface of the lake,[3] call to mind another Service poem, "The Cremation of Sam McGee" and the "strange things done 'neath the midnight sun / By the men who moil for gold."[4] With the damming of the Columbia, useful agricultural land, in short supply in British Columbia, was made waste, and water, a part of nature previously beyond price, was sold down the river to a foreign utility. The "strange things done" in pursuit of power left many local people adrift, not only literally bereft of the physical reference points that had organized their lives in the landscape but also outraged by state actions, which offended both their convictions about good stewardship of the land and their sense of the moderation and balance that secured a good life. Even though the residents displaced by megaprojects of the lower Saint John and the St. Lawrence Valleys did not consent to the upheaval that befell them, for a time they could comfort themselves by saying that their sacrifices served national purposes. For the people of the Kootenays, there was no such consolation.

The displaced residents Halleran records speaking of their losses dwell on flood rather than fire, for the humans most unsettled by this change were the people of the Arrow Lakes, who sustained themselves by gardening, logging, and raising animals. Among them were the Spicers, market gardeners who provisioned the region from a sizable acreage of rich soil on the edge of the village of Nakusp. The titles of two contemporary books by Donald Waterfield, a Spicer kinsman, *Continental Waterboy* (1972) and *Land Grab* (1973), catch the mood of the valley in these troubled years, as does a compelling ethnographic study by James Wilson, a Massachusetts Institute of Technology-trained civil engineer and planner employed by British Columbia Hydro to effect the relocation of the valley residents. Wilson called his book *People in the Way* (1973). When *The Reckoning* premiered in Nakusp seven years after the storage reservoir transformed the lake and the valley, the editor of the *Arrow Lakes News* called the treaty a "terrible sell-out to the Americans," a widely shared local view, and ventured the hope that "future governments [would] not be so short sighted and [would] preserve some of this beautiful old world for us."[5] This chapter, through the advice and commentary of a local agricultural leader, considers how the time horizons, the sense of appropriate market scale, and the weight accorded to profit and to sustainability differed between the economy of mixed agriculture and the economy of commercial monoculture in the 1950s. This discussion is framed as an issue in the embodied environmental history of that progress privileging time. And then, because the flowing Arrow Lakes were a central reference point for life in the valley, I examine the challenge that the changes in the Columbia – from river to reservoir, from seasonal flow of melting snow to stark rise and fall of inventories of electricity in waiting – presented to inhabitants who had embodied a habitat fundamentally reordered by the dam. What ties these two discussions thematically is the personal and community turmoil that arose from the commodification of nature and the derogation of sensual awareness. As inhabitants adapted to the stark and massive landscape change, their everyday practices had to accommodate a different reckoning between use values and exchange values with regard to both their goods and their time.[6]

The people displaced by the storage reservoir of the Columbia were not the first residents of the Arrow Lakes in historical times. Before them, and for a time cohabiting among them, were the Sinixt, Interior Salish peoples known in the recent anthropological literature as the Lakes people. In 1811, David Thompson reported on the presence of the Sinixt and their staunch habits of maintaining their distance from developing European commerce. Into the twentieth century, one of their villages persisted at Kuskanax, on the northern boundary of Nakusp; they fished and hunted caribou there by Kuskanax Creek and in the narrows between the Upper and Lower Arrow Lakes at Burton. Since European contact, the Sinixt had their own history of displacement, first during the 1850s gold rush at Pend d'Oreille River, when they were forced north into the Arrow Lakes region from their territory around Fort Colville, in Washington State, and later east into the secluded Slocan Valley in the 1880s and 1890s as silver miners surged into the galena ore shelves of the Lardeau River and the towns between Kaslo and Nakusp.[7] Like the Arrow Lakes settlers who gradually succeeded them in the early twentieth century, the Sinixt were misunderstood and marginalized in the official classifications of their day, mistakenly grouped with the Kutenia by anthropologists working east of the Rockies and as Shuswap by those working west, their claim overlooked

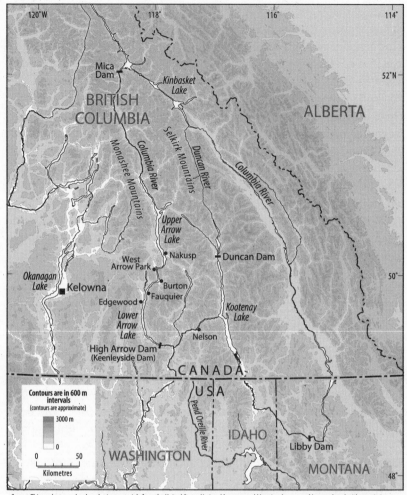

FIGURE 5.1   Arrow Lakes showing dams, settlements, contour lines, and drainage | Cartography Section, University of Western Ontario, Department of Geography

in the sorting out of more honoured contending twentieth-century national interests.[8] By the mid-twentieth century, most Sinixt were officially classified as part of the Colville Confederated Tribes south of the border, and when Annie Joseph, the last member on Canadian rolls, died in 1956, the Arrow Lakes Indian Band was declared extinct,[9] its presence extirpated from the official reckoning of its territory north of the forty-ninth parallel.

The Sinixt knew the Arrow Lakes specifically in time and space, as a sequence of places frequented seasonally by fish and game. These were physical sites of physical activities upon which they physically depended annually as hunters and gatherers. These were also ancestral burial places to which they were tied by continuing spiritual obligations. For the agriculturalists who began to reconnoitre the valley in the 1890s, these lands and waters were also physical realms of use. The littoral grazing places enriched by silting spring floods, the south-facing slopes that would shelter orchards, and the deep peaty yet mineral-rich soils deposited where ancient creeks had met the Arrow Lakes were opportunities, material contexts to be shaped by study and experience into a means of living and a way of life. Once the SS *Minto* was launched in 1898, goods and people began to move more readily into and out of the Lakes settlements, but the livestock and produce families raised in the valley principally met the needs of those who supported themselves nearby in logging, mining, and commerce. As the mountains had sheltered the Sinixt from the demands of the fur trade, so they offered later local agricultural producers a protected market. The extreme attrition of space in the West Kootenays required and rewarded close attention to what the natural endowments of the place would yield and to the quality of life the knitting together of these elements would sustain. For those who dwelt in the valley in the first half of the twentieth century, the profound constraints of the setting as human habitat, the isolation, the steep slopes and deep shadows, the limited arable land – amid natural splendour sustaining a reliable sufficiency – forged an intense sense of presence, of being "here" and "of the place."

Both the Saint John Valley and the St. Lawrence Valley were gateways to the continent, places rich in history, but they were also conduits for Europeans in transit, measuring their North American prospects and moving on. Settlers and stayers in these valleys knew dwelling place as a choice among alternatives, to be defended or conceded according to the currents of North Atlantic politics. Not so in the Kootenays, particularly in the Slocan Valley and the Arrow Lakes. These places were singular for those who lived there and embodied the place, not *a* world but *the* world.

The Arrow Lakes, of course, existed simultaneously in another way, in a strategic realm of possibility, more imaginary than material, more prospective than present. To political representatives of the two nation-states straddled by the Columbia Basin, the West Kootenays was conceptual rather than visceral space, not a "here" to be lived in but part of a systematically featured and bounded "there" to be allocated according to intentions validated elsewhere. Such plans distinguished rather than integrated the possible purposes to which the space might be turned, and they hierarchically ranked potential uses and users. According to one such plan, the Sinixt were made members of the Colville Confederated Tribes in Washington State. According to another, arising to prominence in the mid-twentieth century, the great volumes of snowmelt flowing down from the mountains (one feature of the Columbia River – in terms of annual drainage the largest of any river along the Pacific coast), was taken to define the Columbia Basin as a whole.[10] As a corollary, this schematic of the place simplified the Arrow Lakes,[11] reducing them to a container for the waters of the Columbia River, the vessel in which the river's seasonal flow could be transformed into a biddable aqueous inventory moving in response to distant demands for power. As the villages along the St. Lawrence were remade to accommodate the Seaway transportation route, and the hill farmers south of Gagetown were displaced to make room for NATO tanks, so the improved arable lands of the Arrow Lakes were repurposed as a storage container – scarred by waste and emitting foul odours, but dependable – for electricity in waiting.

Nakusp arose in 1892 as a provisioning and transport centre for the silver mines in the Valhalla and Slocan ranges to the south and east. By 1895, the village was connected inland to the mines by a railway, and by steamers plying the lake north to Revelstoke, on the main Canadian Pacific line. A sawmill producing railway ties and lumber for building prospered on the waterfront. The sand and gravel that underlay the village was unpropitious for cultivation, but by 1897 a laundryman named Sam Henry was raising vegetables for the steamers on a plot near his shop. Alive to the possibilities of the growing market and locally unimpeded by provincial statutes forbidding the Chinese to own land, in 1904 Sam Henry purchased 160 acres (42.9 hectares) of spring-fed rich black soil running back from the lakeshore on the southern edge of the village. Here he built barns and accommodation for his labour force of twenty, moved a modest well-finished wooden house (that the sawmill owners had built for themselves) onto the village edge of the property as a home for himself and his two wives and children, and prospered as a market gardener. After his death in 1912, his first wife, Yip Shee, continued the enterprise until 1922, when a leader of Sam Henry's Tong arrived from Victoria to help the family settle the complicated estate.[12]

For the next quarter century, the new owners, two Cornish brothers, Clement and Philip Buesnel, used the lands Sam Henry had improved. They established the Bay View Dairy and continued to raise produce for the settlement and the steamers near the front, grazing cows and growing hay and silage for winter feed on the fields above. They built a modern dairy barn with concrete floors, a water-cooled milk house, and silos to store feed.[13] Like the succession of settlers in the lower Saint John Valley, who made rich pasture for Jersey cattle of the hill country that once yielded only scant forage of poor quality, Sam Henry, the Buesnels, and, the Spicers made productive land that without them would have been waste.

Through the years before the First World War and in the 1920s, new settlers, lured from Britain by land companies promising propitious acreage for soft fruit orchards, and European immigrants and Canadian migrants established themselves in "landings" along the Arrow Lakes, often after having spent time as wage workers in the mines and refineries of Rossland, Trail, and Creston in the southern Kootenays. The West Kootenay district had 8,446 inhabitants in 1901, 30,502 in 1921, and 60,060 in 1951. After the silver rush passed, logs and poles rafted down the Lakes formed an export base for the valley. More land was cleared for grazing; shelters were built for dairy cattle, poultry, hogs, and sheep; men worked seasonally in the woods; women tended gardens and cooked in the logging camps. Though the maturing orchards planted in the early twentieth century never fulfilled the land companies' commercial promise as an export (for fruit in the Arrow Lakes ripened and reached distant markets later than did fruit in the Okanagan to the west), this mixed logging and farming regional economy provided a modest sufficiency and a good life for households along the Lakes.[14] In 1948, Chris Spicer, an experienced market gardener from southern England, bought the Buesnel lands and, with his new wife, Donald Waterfield's sister Jean, resumed Sam Henry's role.[15]

We know a good deal about agricultural philosophy and practice in the region through Waterfield's writings. He was a leader in the Arrow Lakes Farmers' Institute and, as R. Ronson, wrote regularly from 1951 to 1956 for the *Family Herald and Weekly Star*, a newspaper read nationally in farm households. Waterfield came from a gentry family with traditions in the British and Indian Civil Service. After completing an English degree in agriculture and a term as secretary to the governor general in New Zealand, his father, Horace, settled his family in 1912 on twenty-five acres at Crescent Bay, just under five kilometres south of Nakusp. There, following the recommended practice of the day, he established a twenty-acre (eight-hectare) orchard. Horace Waterfield was not long in the Arrow Lakes. He joined the local regiment in 1915 and died in battle in France, leaving his widow, Elspeth, with three children, the orchard, and a small pension. All three young Waterfields attended the University of British Columbia. Jean completed a degree in biology with high academic standing. Donald, after two years of agricultural studies, embarked upon a farm apprenticeship in New Zealand. By the late 1920s, all were back at Crescent Bay. Through the 1930s, Donald, now married to Freda Brown, the daughter of a Birmingham pacifist and free-trader who, for the good of his health, had taken up the property next door in 1923, diversified production on the Waterfield lands, adding pigs, cattle, and sheep. And he nuanced his politics through a deepening relationship with his father-in-law. Waterfield, who had fractured his thigh in a farm accident in 1928, lost his leg in 1948 to osteomyelitis. His effectiveness on the land, and particularly on fruit trees, thus limited, he turned to agricultural journalism to supplement the family income.[16]

Waterfield wrote from a place where farm-making was part of living memory. Around him farms were being carved from densely forested land grants. He and his neighbours who were also grazing cattle knew in their bones the month or two "of exceedingly heavy labour for man and team" necessary to clear each acre of pasture. The rapidly growing surrounding timber was a resource as well as an impediment to their agricultural projects. Cash in the first two generations came principally from local cedars that were merchantable as utility poles at forty years and from cottonwoods that were merchantable as pulp at eighteen.[17] These arable holdings, expanded by logging rather than by purchase of additional cleared plots, were best regarded as "works in progress" and legacies to be passed on rather than as marketable commodities, not least because there was no active local market in land.

But Waterfield's practical essays, based on a good agricultural education and wide, well-considered experience, addressed the dilemmas of the scale and style of mixed farming. The mixed farmers and lumberers in the Lower Saint John Valley were exceptional in being near enough to markets and suppliers for their Hampstead Co-op Creamery to succeed. More commonly, outside the cities, in mid-twentieth-century rural Canada, transport and power infrastructures were weak and unreliable. Distant markets dominated by powerful wholesalers made farmers price-takers both as buyers and as sellers. Labour became scarce in rural areas as children stayed longer in school, and young people, at least for a time, moved away. An urgent drive to accumulate things, urban things, posed a glamorous alternative to the long-time horizons and the deeply sensuous moderation of rural life. Yet despite the contemporary trend towards large-scale commercial monoculture, both in market and non-market terms, family farms were well adapted to circumstances in many parts of rural Canada in the 1950s.

To this farm audience, Waterfield offered concrete advice couched in an amiable self-deprecating tone, amplified by the jaunty cartoon drawings with which the *Herald* staffers illustrated his longer contributions. For example, he suggested that a farmer shipping only small quantities of cream could get premium prices, yet by using disinfectant rather than steam to clean his milking machines, he could avoid the risks associated with firing up a boiler twice a day in his barn. His plan for a single-storey poultry plant that hugged the ground with insulated laying houses on either side of a feed shed and opened out onto a roofed sun porch and minimized labour, steps, handling, and fuel when keeping fifty to one hundred hens was to be "a small side-line." His instructions for a concrete wood-burning brooder for chicks provided a reliable alternative for those who, in 1951, were still living "off the grid" or whose power lines were prey to heavy winter weather. His review of conventions around the Commonwealth for shearing sheep concluded by reassuring other owners of small flocks that he, too, recovered his facility as a shearer each spring only on the last dozen fleeces, and he offered tips for readers who, like him, considered their "own physical well-being while working, of more importance than saving a few minutes."[18]

He spoke an evasive truth to market power. In the days before marketing boards, only farmers who clubbed together to buy feed in car lots could realize returns on large poultry flocks. By contrast, for those willing to adopt a selective stance of refusal, "a dozen hens kept only to satisfy the family's egg requirements [would] almost always pay a handsome return on a modest investment." He acknowledged that the diversification, which offered agriculturalists a measure of security and independence, daily posed practical contradictions and moral conundrums. A farmer who kept both cattle and poultry needed to "develop a split personality," be by turns a "good livestock man" sympathetically coddling his critters and a "stony-hearted,

Two variations of woodpile architecture include the leaning-tower-of-Pisa type which defied Newton's law of gravity and the conical heaps looking like tall igloos.

FIGURE 5.2   "The Vanishing Woodpile" | *Family Herald and Weekly Star*, 29 April 1954, 5

neck-wringing poultryman" systematically culling the weak from his flock.[19] He wrote with straight-forward and cheerful confidence, affirming the viability of life as a nimble generalist on a small family farm, by choice or force of circumstance at a distance from the contemporary mainstream. He displayed his foibles to his rural audience not as a confession of personal inadequacy but, rather, as a reassuring sign of their shared capacity to persist resiliently in lives that, though organized to maximize satisfactions rather than profits, were purposeful, sustainable, and worthy.

Unlike his sister Jean and his brother-in-law Chris Spicer, who later farmed organically, Donald Waterfield experimented with the postwar pesticides and pharmaceuticals that might have increased the viability of small-scale agriculture, and he urged his readers to follow suit. He added oats to his poultry rations to combat worms; but, just in case, he also doused his pullets with phenothiazine. In September 1950, he observed that "a little of this wonderful chemical," DDT, sprayed on roosts and nest boxes had eliminated mites and lice entirely from his flocks. And the next month, as he and his readers were preparing to overwinter their sheep, he advised that a DDT solution poured over lambs' backs and worked into the long wool under their throats, while making them "giddy for a few minutes," killed the resident ticks and otherwise did not appear to do harm.[20] His approach to innovation, as to life, was deliberative. When kitchen sink postmortems revealed coryza in his fowls, he accepted a government pathologist's advice to add aureomycin to the mash but acknowledged that a better long-term solution would have been to foreswear the risky short-term economy of introducing a neighbour's cocks into his flocks. He attended closely to the qualities of local soils, renewing his orchards with a variety of strains rather than following the provincial marketing organization's advice to invest heavily in the latest "super fruit." He also paid attention to scientific research in biogeochemistry, which showed that a more accurate guide to the nutrients available for plants and animals came from analyzing the nearby vegetation rather than the soil.[21] Above all, he honoured and counselled watchful moderation in the tending of creatures and the land, attending to costs and benefits the market could or did not reckon, and making explicit knowledge and sensibilities that, so long as they remained tacit, were discounted by those who dismissed rural life as idiocy.

His accounts of this tending recall creative workers' descriptions of "flow,"[22] a fusion of sensuous awareness and tactile skill. Sometimes his essays ironically punctured rural nostalgia. His 1954 disquisition "The Vanishing Woodpile" "gilded" wood splitting "with a glamour it never may have possessed." He presented it for strong men as "a splendid opportunity for the physical expression of their muscled ego," a "momentary challenge to young men feeling their oats," the arc of a double-bitted axe a source of aesthetic pleasure now lost to the abominable, "dangerous, stinking and ... maddeningly noisy" power saw.[23] More often, his columns were specific accounts of tactile signs: how to discern whether a bird had been laying by a finger measurement of the width between her pin bones; careful catalogues of temporally specific visual and olfactory sensations, the colour and tensility of which distinguished good hay from feed that would be "unpalatable, barely nourishing ... and potentially dangerous"; and detailed instructions on the sequence of postures, pressures, and the studied balance between waiting and action required, for example, to deliver an awkwardly presenting lamb.[24] Would any of this content qualify as "news" to *Family Herald* readers? Perhaps not. But in one sense it would. For his playful writings were stout-hearted affirmations of the embodied skills and habits of sensuous discernment that made daily life on the land a satisfying modern challenge. By their careful recounting and manifest respect for this labour, his columns offered farm readers alternatives to common contemporary portraits of small-scale mixed agriculture as a dull, vacant, and waning anachronism.

Not that Waterfield's writing lacked a critical perspective on the farm
life of his day or an admiration for the achievements of organized industrial workers. While unionists had bargained for holidays with pay and
shorter hours as mechanization increased their productivity, and as they
sensibly took time to enjoy the fruits of their labour, farmers "appeared
to believe that there [was] inherent virtue in work itself" and drove themselves and their "miraculous new machinery" to the limit. Playing the
contrarian, inviting "adverse comment," and perhaps showing the influence of his own gentry rearing and his father-in-law's socialism, he offered
the "practically revolutionary" suggestion that a farmer "should actually
down tools and rest for the whole of a Monday afternoon." Who could
disagree? "Work we must, if we would eat; but surely there should be
moderation even in work." And so he offered his readers this "regrettably
sententious" (and unabashedly sensualist) resolution for the New Year,
1952:

> Resolved that I, a farmer, will in the future and regardless of any decreasing
> material or financial reward, work less hard and for fewer hours; may I lift
> my nose from the grindstone long enough to see and to hear, to smell and
> to touch, to appreciate and to enjoy, and to give thanks for all my very
> considerable blessings; blessings both tangible and intangible, visible and
> invisible, material and spiritual.

Even the devoutly observant agriculturalist (perhaps especially the devoutly observant agriculturalist), daily immersed in nature's works, knew
that the good steward refused excess. The *Herald* illustrators ruefully accompanied this essay, entitled "Six Days Shalt Thou Labor," with a cartoon
of a man in coveralls resting under a tree, fishing and reading poetry, while
nearby his livestock and poultry bawled in alarm.[25]

Years before Waterfield learned that his valley was to be overtaken by
an extreme instance of modern gigantism, he was preoccupied by the
ethics of sustainability. In this he was not alone. After the war, Canadian
fiscal and social planners and consumer advocates, supported by a sympathetic public, had implemented public polices that redistributed income
and laid the foundations for medicare.[26] Protections for the domestic
market in manufactured goods sustained these strong welfare state policies
in both Canada and Europe, but internationally agriculturalists sold into
world markets in the thrall of "unwieldy and unsatisfactory" "remedies
for gluts south of the border." His response to the 1956 Royal Commission

"Six Days Shalt Thou Labor"

By R. Ronson

FIGURE 5.3 "Six Days Shalt Thou Labor" | *Family Herald and Weekly Star,* 8 November 1951, 5

on Canada's Economic Prospects was to call attention to the crisis of over-production then driving down farm incomes while "glutting the continent with [an] unwanted" production of meat, milk, and grain. He asked, "Have we become so confused by the complexity of modern economics that we can no longer distinguish between good and bad?" Are we unable to grasp the "aesthetic value" of working to live rather than working to produce, when "beyond the acquisition of bare necessities, we all have food, clothes and shelter and, most of us, entertainment as well"? The essay struck a chord and prompted letters of support from agriculturalists across the country.[27]

Waterfield's last major essay as R. Ronson, "The Vanishing Farm," was a portrait of the alternative way he and many of his Arrow Lakes neighbours lived. Some were leaving the land because farm commodity prices had failed to keep up with industrial wages near Nakusp, with wages in logging and the mills associated with newly consolidated timber licences. The farms that remained became more specialized and mechanized. Economics compelled these changes, but specialization made life "dreary" and "anxious," and the man who yielded too thoroughly to mere financial incentives impoverished himself and his family. Indeed, for many the possibility of large-scale monoculture was moot. The physical conformation of their farms alone made hundred-acre (forty-and-a-half-hectare) fields of any single crop untenable for many in the Arrow Lakes, in most of British Columbia, and in much of rural Canada. By 1956, Waterfield himself had enough cleared land and sufficient access to transport to specialize, and his range of production did narrow. But his essay emphasized the necessity rather than the sufficiency of this adaptation. He insisted,

> How can you have a farm and children, and not have horses too? And what is a dinner table without fresh fruit, fresh vegetables and masses of sweet fresh cream? April without sheep is just another month, and with newborn lambs it is Spring. How can you have a barnyard without grunting hogs to slop, or a pasture without cattle fattening?

For the first time, in this last essay Waterfield included family photos, each captioned to emphasize the deficiencies in a mere livelihood for what he counted crucial to a good life. Beneath a picture of his wife Freda, with the family dog and a young donkey, he affirmed, "These animals may be uneconomic on a family farm but they add interest and excitement to daily living." Below a vintage image of his children, Barbara and Nigel, on horseback and clad in work clothes, he insisted that the team had "very little to do, but they help train the youngsters." The "fascination" and "fun" of tending familiar creatures, the deeply sensuous moderation of "working, playing and relaxing" on the land, of learning, knowing, and being intimately in place, made "a mixed farm – with all its hard work, squalor and isolation ... a pretty good place to live."[28]

These essays were not a rejection of modern scientific agriculture. Waterfield had a good formal agricultural education and was an early, if selective, adopter of innovations to raise yields in his fields and to improve the health and productivity of his stock. Nor were they romantic laments for an untenable and anachronistic way of life. What Waterfield rejected, with supple humour and broad irony, was the rising orthodoxy that only places suited to capitalist monoculture could efficiently produce food; that the lands of hinterland small holders, who were providing local markets with fresher meat, fruit, and vegetables at lower transport costs, were best abandoned or turned to other purposes. He risibly refused the contemporary confusion of material accumulation with sensual delight, and of bigger returns with a better life, because – though anyone, almost anywhere, might on occasion tritely affirm that "money can't buy happiness" – in the Arrow Lakes, and perhaps in many of the places where his readers lived, pleasures were often muted rather than amplified by the demands of the market.

In 1959, a few days before Christmas, Waterfield telephoned his brother-in-law Chris Spicer. As the story has come down through Janet Spicer, Chris' daughter, her uncle Donald told her father first to sit down and then said, "Chris, you're going to lose your land." By 1959, Waterfield had become president of the Nakusp Chamber of Commerce and, in that capacity, had attended a meeting at the base of the Lakes in Castlegar, where the engineering maps for the proposed dams for water storage and power on the Columbia River were displayed. At that time there seemed to be two options in contention for remaking the Canadian portion of the Columbia Basin, one for a low dam at Murphy Creek, which would back up the waters of the Arrow Lakes only to the current high water mark of 430.07 metres above sea level and hold it there; the other, more ambitious, was for a high dam below Syringa Creek, which would establish a new high water level at 442 metres, provide two and a half times more storage capacity, and in the process flood out 1,700 Arrow Lakes inhabitants, a third of Nakusp, all the settlements to the south, and almost all the arable land in the valley. At full pool, the reservoir behind the High Arrow Dam would inundate parts of Chris Spicer's fields. At this level, the waters of the lake would be lapping at the front steps of his home.

The representatives at the Castlegar meeting counselled against any quick action, featuring themselves as trustees for the environment of the Columbia Basin and its usefulness read large, with "no moral right to sell or destroy land, no matter what the monetary value might be." The decision of the Nakusp Chamber on 21 December 1959 was to oppose the High Arrow Dam. They too were thinking long term, aware of developments in nuclear generation just south of the border at Hanford, in Washington State, and in central Canada at Chalk River, concerned that – while the land that nourished them and their neighbours would be irretrievably despoiled by the reservoir – "in 50 years time atomic power may nullify the need for electrical power produced from water stored in the Arrow Lakes."[29]

What they did not know was that the relevant Hobson's choice was not between nuclear and hydro, or between a low or high dam at the base of the Arrow Lakes, but, rather, as Matthew Evenden has established, between fish and power, for a Low Arrow dam would have entailed a diversion of the Columbia's waters into the Kootenay and Fraser Rivers, disrupting the latter's valued fishery. Nor did they know that in distant Ottawa, three years before the engineering plans were displayed in the valley, the political decision had already been made. The commercial salmon runs of the

Fraser River, debouching into the Pacific near Vancouver, would be saved from the perils of dams and slack water, while the Columbia Basin would be the "power generation region," the specified site for the "displaced environmental impacts" of dams and storage reservoirs. Thus, Janet Spicer's truncated and apocalyptic recollection of her uncle's conversation with her father caught the truth of the matter. As conceptual space, the Columbia River had already been reclassified. To this internationally determined resource strategy, the habitat along the Columbia would be sacrificed just as the pastures and timber stands of the lower Saint John Valley had been forfeited to NATO military preparedness.[30]

Donald Worster tellingly situates the turn of mind that informed these postwar state decisions. In the corridors of political power, development came to be accepted "as a single cultural standard against (which) all people could be measured" and "a formula to be applied everywhere." Those in the region who counselled moderation rather than "unlimited materialism and accumulation" as a way to live a full life on the land, and who honoured habits of trusteeship rather than "exploitative attitudes toward nature," thereafter had little influence over the policy agenda and little control over their dwelling places. This was the case in the St. Lawrence River Valley and in the Saint John River Valley and, until three decades later, when the Columbia Treaty was renegotiated, would be the case in the Arrow Lakes as well.[31] Thus, the lines of "The Reckoning" are telling: things would be done "that you had no right to do," and, for many years, the residents of the Arrow Lakes, more or less on their own, would be left to "foot the bill."

The political, diplomatic, and legal history of the Columbia Treaty has been closely recounted by others, and work on the ecological effects of the storage reservoirs on non-human inhabitants of the Columbia Basin has begun.[32] Instead, I want to consider the effects of the Columbia Treaty on the habitat as it was embodied by human residents of the Arrow Lakes. Previous chapters have explored how inhabitants weathered the challenges to their senses of competence, safety, certainty, and autonomy, all of which were affected by radical changes brought to the places where they worked and lived. The focus in this instance is on the sense of time.

For Arrow Lakes residents, the waters of the Columbia were a central "reference point," an internalized "compass bearing"[33] through which their sense of direction and their perceptions of distance and depth were unconsciously and corporeally merged. Before the dams were built, the continuities in time passing were marked by the steadily moving river, its changes heralding the different rhythms that each season brought for work and pleasure in the valley. With the dams, the waters changed. They became commodities. They moved as commodities move, in response to demand. The qualities of this transformation usefully recall the work of historians Edward Thompson and Herbert Gutman on the transitions from artisanal time to industrial time. Attending to the reconstitution in the "materiality, corporeality and political economy of bodies"[34] that this remaking of the environment required also situates the shared predicament of the valley residents within the psychic space philosopher and literary critic Julia Kristeva conceptualizes as abjection. This elaboration helps make accessible an elusive and viscerally persistent strand in the Arrow Lakes story, a strand that is also threaded through the embodied narratives of Gagetown and the Seaway towns: the abasing challenges to self that inhabitants met as they attempted to make sense of the habitats that others had substituted for the places they had come to know, by honed reflex and habit, as extensions of themselves.[35]

"We know more than we can tell."[36] Bodily, we unconsciously depend upon habit and reflex to make our way. Even dwellers in urban built environments take their settings into themselves in this way, knowing, without being able to specify, the height of the steps they climb daily on their way to work and what it is about the sound of the furnace in their homes that makes them confident at bedtime that they will be warm until morning. We all seamlessly incorporate such sensuous regularities into our routines and assume these architectural and technological patterns to be the norm. Without forming thoughts about their enduring nature, we behave as if they will persist.

If this is how urban dwellers make the material artifices of their lives ordinary, perhaps those who live in environments less apparently altered by human intervention naturalize the places they know even more. Just as the shepherds University of Adelaide anthropologist John Gray has described in the Scottish borderlands learn the hills, so do people who live with a river learn its patterns and rhythms. Recall the Iroquois boat builders and fishers for whom the turbulent St. Lawrence, once studied, became not treacherous but interesting. Through specific practised bodily modes of attention and stores of visceral knowledge that unconsciously merge self with dwelling place, inhabitants become competent in place, able to keep themselves safe and fed and to find pleasure in their surroundings. Donald Worster presents one part of this process as a conscious, deliberative task. People learn to "think like a river," to know it as a "second nature,"[37] as an autonomous and independent source of energy, and to systematically discern and categorize its otherness.

But in tandem is the other learning through practice by which inhabitants sensuously engage in their habitat and assimilate their surroundings somatically as the site of their being. They know the features of their habitat as they know themselves, not as other. Twenty years ago, James Wilson, the BC Hydro planner, observed how the contours and features of the shoreline, the flora and fauna, "the slant of the sun, morning and evening, the wind on the waters and the waves slapping on shore" had been "woven over time into the daily living patterns of those who dwelled among them and had assimilated them, we know not how, into their psyches."[38] Thus does the distinction between nature and Worster's second nature become both consciously elusive and germane.

The inhabitants of the shore and those who worked on the Columbia had learned what the river was through that particular "autonomous and independent" other. They took the rhythms of the Columbia to be *what moving water was and did.* Beyond conscious thought, they took their embodied knowledge to be about the primordial character of flowing water, about traits that inhered irreducibly in its nature. The succeeding deep rupture in the ontology of their dwelling place, the unmaking of the nature they had embodied, is key to understanding what early students of the effects of large dams observed globally[39] – "physiological, psychological and socio-cultural stress" among the displaced. How, with no practical experience of the difference, could inhabitants have grasped that after the dam made the reservoir, the ordinary regularities they knew, through the same processes by which they knew themselves, could and would be undone? How could they grasp that the very nature of the river would be gone, leaving them to live in "an engineer's dream," governed temporally by market reasoning, its chronology forged in a commodity calculus?[40]

By 1961, the raw dimensions of the impending change in the West Kootenays could be clearly summarized for the national press by the local member of Parliament, Bert Herridge: "all the farm settlements along the lakes; all the pole yards, wharfs, beaches, homes, ranches, 18 communities along a 158 mile stretch, and 1,600 to 1,700 people" would be influenced by the flooding.[41] These raw dimensions did not hint at the kinaesthetic and temporal transformation ahead: "The essence of a river is that it flows."[42] People of the Arrow Lakes knew the movements of the Columbia as the cycles of "God's Water,"[43] an "autonomous, independent" other whose rhythms had become the pulse of their daily lives. "God's Water" moved according to nature's time.

Why attend particularly to movement? The human sensing body is an integrating, not a segmenting instrument. The perception of movement relies simultaneously upon sight, touch, and sound, situating them among multiple experiences of being in place, merging deep stores of past learning about directionality, duration, distance, and depth.[44] Their valley was narrow, and the bodily knowledge of the movement of the wind, birds, and animals, and especially of water, was foundational to the way the people of the Arrow Lakes knew their habitat, to their processes of emplacement, their understandings of self in place. One of the first published plaints against the planned reservoir began by considering movement. Attempting to visualize "what it would be like if the High Arrow goes in and the water comes up," Robert Roder pictured the "kids" at play in a deep rapid death trap. He foresaw "much tragedy and bereavement and heartache" and named the change as a trespass against nature: "They want to commercialize and make profit out of every good gift God has given for the benefit of all."[45]

People of the valley knew the moving river by bodily encounters: "You miss the lake because it was there and *near* to you." "The lake had beautiful beaches because of that nice tide and it *kept everything washed*."[46] Their own seasonal rhythms were in time with the changes in the shoreline made by the cycles of the river's movement and the attendant changes in the depth, temperature, and speed of the water. They knew, as nature/second nature, when certain beaches emerged, when sandbars became available to water cattle, when sheltering points and islands made the water comfortable and safe for swimming. Their competence as providers depended on a series of honed tacit assumptions about how the moving wind altered the moving water, how the volume of the flow changed the depth and the currents in narrows through which logs must pass, how to use the lands yearly enriched by silt deposited by the spring floods, and when the fish would be in the steams and the deer browsing by the shore.[47]

Their sense of their growing children and of their livelihood was one with the cycles of the river. This is Ernie Roberts, whose community of West Arrow Park was extirpated before the inundation for reasons that are now difficult to fathom:[48]

I had a logging sale right above my house and every day I'd go up and cut trees off the stump, skid them out with horses, put them on the sloop. Sloop them to the lake and get them down to our place. Then our oldest boy had his seat by that time and he was standing by the gate waiting for me to get

FIGURE 5.4   Two photos taken by Val Morton: (left) stacking bales of canary grass harvested from the flood plain; (right) a self-feeding silo of Val's own design | Morton family photo

on that load of logs and ride down to the beach with me and I would just roll them off at the beach. And then we'd scale them in the spring.

Before the High Arrow Dam, the wharves along the shore were the major human alteration at the meeting of water and land. These wharves were essential until 1954, for the villages were provisioned by the *Minto*, which paused wharf by wharf along the lake, and were important thereafter for loggers, fishers, and for travel by water (valley roads were very poor). Before the inundation, the seasonal changes in the depth and flow of the Columbia and the range between high and low water, predictable from past experience, were built into these structures of human occupance, materially marked by the height of the wharves: "If you had a bigger boat, they were all in the water at all times ... you could tie it to the wharf there, no problem."[49]

Two years after Robert Roder of Nakusp and his neighbours began to try to visualize the new Arrow Lakes to be made by the Columbia Treaty, Peter Oberlander (a University of British Columbia planner) and H. Hunter (the Hydro Authority lawyer) chose visual signs as the way to inform valley inhabitants of the dimensions of the impending change. In the summer of 1962, surveyors, using yellow-topped stakes and lines of yellow paint, inscribed the height of the reservoir at full pool on the shoreline by four settlements – Castlegar and Revelstoke at the south and north, and Edgewood and Nakusp on the west and east banks of the middle stretch of the Arrow Lakes.[50] Right away Chris Spicer realized the inadequacy of the scopic representation, that the painted line as it passed by his doorstep marked only the reach of "the still pond" and "did not allow for wave lash and erosion."[51] But surface movement was but one relatively small deficiency in what the visible survey stakes could convey about how the moving waters of the Columbia would be altered by the dam and, by extension, how inhabitants' reference points for the orientation of self in place would be disordered.

Inhabitants of the Arrow Lakes understood changes in the water to be tied to the rhythm of the seasons, the nine-metre difference between high water in the spring (from snowmelt off the mountains that surrounded the lakes) and low water in the fall. In the years between the signing of the Columbia Treaty and the flooding of the reservoir, residents worried about the ugly mud flats that would be exposed by the promised twelve-metre increase in the spread between low water and full pool. There is no sign in the local press that residents anticipated that the movement in the water levels would be other than a relatively predictable seasonal cycle, some mediated echo of "God's water" in nature's time.[52]

To be sure, the people of the valley understood themselves to be in the thrall of a colonial relationship to Canadian cities on the BC coast, to the influential industrial power users whose headquarters were there, and to the collective will of US citizens, governments, and corporations expressed through the decisions of the Bonneville Power Authority. The Columbia Treaty had been signed without any consultation with local residents. Only after the fact were they given a hearing, and then they were permitted to speak only about technical aspects of the water licence. Before the damming of the river, inhabitants of the shore railed against their lands being "governed by the caprices of a foreign power," against losing scarce arable lands to provide power for US competitors to Canadian industry, and disproportionately bearing the burden of "trying to improve on nature." Some villagers of Nakusp suggested that, rather than celebrate the opening of the High Arrow Dam, "a piper be hired to play laments along the waterfront" or that the provincial premier and the head of BC Hydro be set adrift on a burning barge while local people celebrate "with song and dance" on shore.[53] But the caprice they protested was the building of hydro facilities without sober consideration of the alternatives, building so pro-digiously ahead of demand, and building without timely local consultation. Only after the reservoir began erratically to rise and unexpectedly to fall did the perturbing possibility begin to emerge that the storage reservoir would not cycle according to the predictable patterns they had experienced and embodied as the nature of water but, rather, according to the priorities of a foreign power.

By visualizing on the basis of their historical experience with low water and contemporary staked and painted survey inscriptions on the ground of the storage basin at full pool, residents recognized that, in many places along the littoral, including at the village of Nakusp, the distance between the new low water line and the new high water mark would increase by several tens of metres. Their "beautiful lake" would have an "abscess all around the outside" of sand, mud, and debris.[54] Between the signing of the Columbia Treaty and the inundation, inhabitants successfully organ-ized on two fronts to mitigate effects of this anticipated change. Residents demanded and got Hydro to pay for a massive clearing of the area that would be alternately drowned and exposed by the new vertical cycle in the river as well as for a 7:1 sloping waterfront, stabilized by soil-cement and topped by a walkway and park, for Nakusp. And yet, as the editor of the local newspaper affirmed after Hydro had done as much as it was prepared to do in mitigation, the "changes in the valley [were] hard to accept." In the summer of 1967, a time in which most Canadians were celebrating that their nation had made it through its first one hundred years, inhabitants of the Arrow Lakes were fearful. The dimensions of the present and impending changes left them estranged from the dwelling place through which they had learned about themselves and their world. The "slashing and clearing along the shoreline" gave "a bald appearance to the landscape," so that the once "'cosy' ... tree lined highway" now gave residents "the feeling of having arrived at the 'edge of the world' with nothing much to keep them falling off." The new waterfront was "tidy and neat" but no longer sheltered by the now denuded "Point" to the north that, "since time began, [had] done much to protect the town from the storms of the lake." "The tremendous physical changes taking place in the area" disoriented those whose habits of dwelling and living had been formed in daily reciprocity with the river, so that Nakusp no longer felt like the "safe place" the Sinixt reportedly had named it.[55] The visual signs had given warning of the spatial changes in the river, and these were hard enough

FIGURE 5.5   Front at Nakusp, looking northeast during a drawdown, with the Spicer rose garden at top. The rippled area just below is the soil is cement installed to stabilize the slope. A floating dock and houseboats are in the foreground. | Arrow Lakes Historical Society

to absorb into the habits of people accustomed to daily contact with its waters.

They had oriented themselves, located themselves by direction, duration, distance, and depth along the shore and on the water, by learning how the river moved in time as well as space. Their experience with the sound, the temperature, and the speed of the water, with how it behaved seasonally in fair and foul weather, all tied to temporal changes in the flow, organized their work and their recreation, their senses of competence and pleasure in place. As soon as the High Arrow was closed in the spring of 1969, the waters in the reservoir began to come up at what seemed "an alarming rate."[56] Then and thereafter, both the rise and the drawdown in the water were precipitate and disorderly by comparison with the movements residents had learned as the "nature of water" in their home place. Like artisans forced from the workshop time of the task to the factory time of the clock, they moved from an environment whose constraints they knew and whose ordinary limits they, through experience, had learned to accommodate, to a mechanized setting whose pace was governed by a calculating proprietor. In the case of the Columbia reservoir, these limits, while contractually defined, were enigmatically prospective and operationally inscrutable.

The first winter there was little snow and when the full drawdown occurred in January, the reservoir at Nakusp became "an ugly mess of twisted, black stumps protruding from ugly mud flats." Villagers longed for a cover of snow. The rapidity of the drawdown made people fear (justifiably, as later experience proved) for their wharves and boats.[57] The next spring the drawdown was greater and occurred earlier, suggesting – as Denis Stanley, editor of the local paper, hypothesized using diction that bespoke multiple layers of powerlessness – that "BC Hydro and its puppeteer American friends [were] anticipating an early and heavy spring run-off."[58] In 1972, the reservoir did not begin to rise until 12 May. New volumes of cold water continued slowly to flow into the lake until early July. Only then did the level stabilize and the water begin to warm for swimming. The next year, at peak holiday season when residents expected both to be able to use the lake and to earn some income from the tourist trade, the flood gates of the dam were open wide, the residents left with no water "to cover our barrenness." The tourists just plain left. Through July, the drawdown continued at a rate of fifteen vertical centimetres per day so that by 1 August no safe beach remained in Nakusp. BC Hydro did not seem to be authoritative, or at least consistent, in its statements about the reservoir. At mid-month the chairman of BC Hydro suggested that perhaps three or four times in a century the waters would not rise to full pool. The next week, citing correspondence with Hydro officials, Stanley reported that about one year in five, the reservoir would not remain full through the summer. The Recreation Commission cancelled the swimming program for village children, and villagers were warned to take great care entering the water.[59] Winds created sandstorms along the barren stretch of shoreline that once had been the settlement of Arrow Park and the site of Ernie Robert's home. The lower the waters drew down, the more of the remains of the villagers' houses, barns, orchards, and pastures were exposed as haunting reminders. The wide expanse of the Spicers' once-flourishing market garden emerged on the edge of the village as a tangle of grass and weeds. The next year, contrarily, the reservoir rose to 440.74 metres, sixty-one centimetres higher than specified in the Columbia Treaty, bringing the Bonneville Power Authority a considerable excess volume of water for sale, while valley dwellers worried at nights as summer storms lashed the water of the lake to unaccustomed heights.[60]

FIGURE 5.6   In the top photo, Chris and Jean Spicer work in their market garden (pre-1968), with lake, log booms, and Monashee Mountains behind. The bottom photo shows the market garden after the reservoir replaced the lakes. | Arrow Lakes Historical Society (top); Joy Parr (bottom)

Through the early 1970s, as the topography of the bottom of the lake proved unrecognizable and the movement of the water unpredictable, the most seasoned navigators of the valley were repeatedly abashed. The ferrymen at Galena Bay, to the north, sank their vessel a couple hundred metres from shore, forcing their passengers into life rafts, and their boat was out of service for days. Bill Barrow, a skilled and cautious man who had plied the waters for decades, tore a hole in the two-and-a-half-centimetre thick hull of his tug, the *Canyon II*, on a stump the Hydro workers had not removed. He was left with repair bills and unmet contracts, and his two crew members were without work for the duration of that season.[61] Ernie Orr moored his boat for two days and returned to discover it high and dry, the fibreglass bottom cracked because "they" had dropped the reservoir sixty centimetres in his absence.

The waters of the reservoir were not the waters the inhabitants of the Arrow Lakes had embodied. Orr concluded, "There's no way they can really control it," which meant that he had lost a measure of control over his own life: "You've got to be watching all the time ... 'cause you never know what they're going to do with this lake."[62] In spring, the lakeside village of Nakusp looked (and looks) "like a nice little community sitting on the edge of a semi-arid desert."[63]

By the summer of 1975, the residents of Nakusp, "short of blowing up the Arrow Dam at Castlegar and declaring war on the USA," a notion without much staying power before the realities of continental politics, knew what they wanted: that "the reservoir be filled in May and be held full until the first snow," so that more of the beauty, usefulness, and predictability they had known in the time of "God's water" would be restored to them. And they knew, on the basis of fifteen years' co-habitation with the Columbia Treaty, that even their less explosive and bellicose fantasy, that the reservoir be near full from May to September, would not come to pass. Their clamour raised from BC Hydro's Vancouver office only a press release noting that the reservoir could be held at full pool each year for the last two weeks of July. This was all that the operation of the reservoir "in accordance with the Columbia River Treaty" would allow.[64]

Denis Stanley, the editor of the *Arrow Lakes News*, portrayed the villagers as having been "debased" by "the brunt of a colossal blunder." The diction he employed to describe the alterations in his dwelling place are evocative: the pall of sediment rising from the mud flats, so thick that residents on the hills above could not see the village; the dust and sand that permeated homes; the loss of contact with the water; the loss of yards of topsoil "blown to oblivion"; visitors moved to tears as they crossed the lake on the ferry at Needles and began to make their way through the desolation between Fauquier and Nakusp.[65]

Stanley described himself and his neighbours as inhabitants of a shared and brutally altered dwelling place. He described his neighbours as without a recognizable external foundation upon which to know themselves in place. Stanley emphasized that the pall of dust occluding the village from view erased borders between parts of the landscape and between self and other. The valley had become a place where sediment, caught by the wind, permeated all dwelling places. With the environmental feature that had secured their understanding of time and place compromised, people's sense of self was unsettled. The boundaries between self and other became uncertain.[66]

The first histories of the modern displacements made by megaprojects, centrally among them large dams, have rightly emphasized property losses, the struggles to resist expropriation, secure restitution, physically relocate and rebuild – elements of these transformations keenly felt and readily accessible to historians through texts: the petitions, commissions, and legal actions arising from physical transfers of proprietorship. But as Wilson noted in his book about the Columbia and as Goldsmith found in his transnational study (both contemporary accounts), by disordering the material reference points for inhabitants' relationships to their habitat, these large engineering works had sundered the ways place had been "woven" into "their psyches," their selves in place.[67] Emplacement was a bodily process, a learning lodged in muscular habits and neural reflexes that made the rhythms of habitat the rhythms of self – by extension incorporating the reckonings of distance, depth, duration, and direction – and situated people in their habitat.

You'll recall that the Enlightenment philosopher Descartes, insistent upon the consciously conceptual foundations of self, affirmed, "I think, therefore I am." For the valley residents, left to "foot the bill," the anchors of identity were sensual as well. Donald Waterfield and his neighbours knew themselves through the places they had embodied: "I am here, therefore I am."[68] For the people of the Arrow Lakes, this postulate entrained the disturbing corollary question, "Who am I when this here that is me is gone?"

Let us conclude then, by returning to the part of the valley where we began, near the Sinixt fishing place at Nakusp, the place that from 1904 was Sam Henry's market garden, that from 1922 was the Buesnel's Bay

View Dairy, that from 1948 was the Spicer farm – the celebrated acreage of rich black soil, sub-irrigated by a warm stream, fronting the lake with a southwest exposure on the southern limits of the village, its microclimate protected by an escarpment circling behind – the place where the yellow stakes marking the reservoir at full pool were first sighted in town. The Spicer lands were as rich as the best in the valley. The family was also more amply endowed than most with cultural capital. In these senses, their story is only their own. But the Spicers' losses came to stand for the challenges faced by the 1,700 displaced residents of the Arrow Lakes. The tactics they employed to resist, mitigate, and persist engaged their neighbours viscerally as well as symbolically as their fields had been the region's main source of fresh produce and Jean's celebrated skills in floriculture had made the beds, borders, and bowers of her garden a highlight of visits to town. Their response to abjection was to build a new "here," turning their losses into personal possibilities and shareable civic improvements.

Jean Spicer spent the 1930s with her mother, living at Assart, the house on a height of the Waterfield land, cooking for her sister Nancy in the Bluebird Cafe on the main street of Nakusp. She continued to explore the region, especially its high alpine reaches, learning by systematic botanizing, and recording what she found in photographs and paintings. Her hobby, begun in the Depression years, was collecting dog hair, from as far afield as Virginia and Kansas. Barbara spun this fibre into yarn and Jean knit it into sweaters and socks. In the summer of 1941, amid wartime shortages, the Vancouver *Province* reported on this hobby: the touring columnist called it "a profitable enterprise."[69] Their mother limited her daughters' social circle to the other gentry families nearby, selected parishioners of the Anglican Church, and the "boys from the bank" who were invited to Saturday night dances at Assart.[70] In 1948, at age thirty-nine, Jean's romantic fortunes turned.

Chris Spicer was thirty-five when he arrived in Nakusp that year. He came from Sussex in England, had been educated at Eton and the Wye Agricultural College, and, in his twenties, ran a large market garden in Hampshire. By all accounts he was an exceedingly gentle man, but after one of his brothers was lost in battle, he signed up and piloted a Wellington bomber through forty-nine missions.[71] After the war, with limited funds he came to Canada, looking for land. He began in Newfoundland, where a brother was practising medicine, and worked his way across the country. He told journalist Stuart McLean that he came to Nakusp to work on a miserable mink farm, met Jean at the Bluebird Cafe, and found the Buesnel brothers so eager to retire they were willing to take what he could pay for their land and let him work off the balance.[72] He married Jean the next year, and in 1950 their twin daughters, Janet and Crystal, were born.

Whereas Donald Waterfield's approach to the use of postwar agricultural chemicals was selective, the Spicers' was not. Chris told McLean he'd moved on from Ontario in 1947, having found there "far too many chemicals" for his liking.[73] Observant, skilled, working with the natural endowments of the site, and protected by the physical barriers around their fields, the Spicers in time achieved high levels of productivity without pesticides or herbicides. Through the 1950s, their sales in the local market quadrupled. The quality of their produce, as displayed on the show bench (the whole family were avid horticultural competitors),[74] did not seem to have suffered as a result. Early on, in 1956 and 1957, they swept the agricultural classes at the West Kootenay Exhibition in Nelson.[75] After 1962, influenced by Rachel Carson, they also became concerned about the effects of retail lighting and packaging on the nutrient value of their vegetables, and urged the vendors they supplied to sell their produce from open bins.[76] In the early 1970s, Chris championed fresh mountain water, made safe by source protection rather than chlorination, as the best way in ensure healthful and palatable drinking water for the village.[77]

They mechanized. By the mid-1950s, they were using a tractor as well as horses to work their fields and move loads, but, like Donald Waterfield, they closely scrutinized the qualities of machine work. Noting that the blades of a baling machine made grasses hard and stubbly, coarser as feed and less soft as bedding, the Spicers stooked fodder and hay for their stock rather than using a baling machine.[78] Granted, this consideration for their animals, consistent with Jean's long-standing animal welfare concerns, was possible given the scale of their holdings.[79] It was also appropriate to the scale of their local market. They pursued the sensual pleasures of their place, climbing, swimming, skiing in season, engaging the market to live while regarding material accumulation with the scepticism common to their class.[80] Both of and not quite of the town, they followed more closely the path of Jean's patrician mother than that of her populist brother. As conflicts over woods practices heightened in the Kootenays in the 1980s, Jean's public statement against "making profits out of wild animals or plants"[81] would have bemused the forestry workers and guides among her neighbours.

The Spicers' most productive ten acres (four hectares) were flooded by the reservoir, and the black muck that had fed the vegetables that had fed the town was gradually carried off by the waters. As BC Hydro's consulting engineer predicted in 1966, their fields became a slough. But it remained *their* slough. Because their property included an accessible bench on the best route into town from the south, some of which they had sold years before for road and power allowances, exceptionally in the valley their holdings had a market history and thus a price. Using these precedents, a good lawyer, and patience, Chris Spicer mitigated his losses. In the end, for a flow easement over the lower fields alone, Hydro had to pay him twice the price he had demanded early on for the property free and clear.[82] The family retained title, could mine muck from below the flood line to improve the small field above near their house and barns, and could cut hay off the flats in years when the reservoir did not reach full pool. But their search for replacement acreage within their means was unsuccessful, and at age fifty-five and fifty-nine, respectively, their farm-making days began again. After the inundation, Chris and his daughter Janet became gypsy market gardeners. By draining and nourishing dispersed pockets of soil (rented and borrowed northeast of Nakusp) and erecting three-metre high wildlife fences around a 107-acre parcel purchased fifty kilometres to the south across the lake at West Arrow Park, they continued to meet a portion of local demand. But with so much of the arable valley land now under water, the Arrow Lakes were increasingly tied to the continental food provisioning system and residents more often had to be satisfied with imported produce.[83] From 1971, Jean tended flowers closer to home as the guiding force in the development of a public promenade and arbour for the waterfront. This was named the Spicer Rose Garden in 1992, the centennial year of the village.[84]

Stories like these come to us from dam sites all over the world: Arundhati Roy's accounts of the displacements in the Narmada Valley in India, a country where 3,300 dams were built during the fifty years after Independence; Mark Fiege and Keith Petersen's histories of farmers, fish, and slack water along the Snake River in Idaho and Washington in the United States; Blaine Harden's biography of the life and death of the Columbia River south of the forty-ninth parallel; Patrick McCully's transnational study of silenced rivers; Eric Swyngedouw's complex portraits of the altered rivers of Spain; and the political and cultural analyses of the epic struggles of the James Bay Cree.[85] What does this embodied environmental history of the Columbia reservoir, recounted through the sensuous learning and practical ethics of the Waterfields and the Spicers and valley residents' shared experience of the changing movement of the river, add to the accumulating store of critical narratives concerning hydroelectric development?

The difference between the excised strategic spaces of continental resource planning and the material places – the landscapes of dwelling and producing – that the plans would destroy emerges forcefully through Waterfield's advice about practice. His arguments for mixed farming treat scale as a multiply presenting possibility rather than as a singular, universally applicable requirement. This is because his appraisal of place, an estimation we might assume his many rural readers shared, was more heterogeneous than the strategic plan would admit, and encompassed qualities of the place the planners of the day would not reckon. There were places, even places conceptualized as markets, where smaller scale made both economic and environmental sense. The outcry along the Arrow Lakes against the destruction of arable land was pragmatic. Forestry would continue to provide employment. After three generations of settlement, services were in place along the Arrow Lakes. The valley was a good place to live. There would be a continuing local market for agricultural production. If only power development could have been deferred until technologies less destructive of other resources presented themselves, these inhabitants could have eaten more of what their neighbours produced (fresh vegetables and fruit, meat, poultry, and dairy) and less of what needed to be transported over great distances by vehicles burning fossil fuels. The vaunted efficiencies of large-scale, chemically enabled monocultures did not reckon this difference. The dams forever foreclosed this more sustainable alternative.

Waterfield's brief in favour of small-scale mixed agriculture tied to local markets and against the alternative of a continental food system extended beyond this argument for a more inclusive reckoning of the costs. He pressed the classical distinction between quality and price. Price represented only exchange value. Too much that was valuable neither could, nor ever would, be offered in trade. Any sensible agriculturalist farmed for profit, but no reasonable person mistook a good livelihood for the whole of a good life. In this he adopted neither a romantic nor a nostalgic position; rather, he subscribed to an essential tenet of the contemporary union movement and the national commitment to a modest and modern welfare state in Canada. The key was to be attentive to sources of satisfaction and sensual delight, which, by their nature, eluded the commodity calculus, and to be mindful of how much was enough and what was too fundamental to ever be fungible. The residents of the Arrow Lakes knew their place as a physical realm of multiple possibilities, a material context of natural splendour that had sustained a reliable sufficiency. To find this protean place attenuated to a reservoir, or, as James Wilson called it in 1992, a giant bathtub ceaselessly being filled and drained,[86] was to be made to foot the bill for the delusion that development was synonymous with improvement.

The industrial reordering of the temporal rhythms of the valley amplified this loss. The development ideologies that underlay the Columbia Treaty took the changes humans effected to be unambiguous signs of progress. Among the residents of the Arrow Lakes, the practical experience of change through time was more mixed. Even though the displaced Sinixt had become forgotten history, passing from the living memory of the place, the abandoned mine heads and cabins, the pastures growing back to alder, the gnarled orchards planted by hopeful settlers who since had moved on, remained as daily evidence that time both made and unmade human plans. Inhabitants' well-being was derived from a welter of activities, some governed by the prices of the season, others organized around the long-time horizons of dwelling and being in place. The waters of the Columbia, rising and falling to the rhythms of the seasons that commonly governed work on the land and in the woods, affirmed this necessary complementarity between nature and second nature. When the reservoir turned the water of the Columbia into a commodity whose movement was governed by demands from distant markets, the whole valley as habitat and the daily lives of its inhabitants were increasingly held hostage to value determined enigmatically at some contractually occluded point of sale. Waterfield's writing, by revealing the stores of sensuous discernment and ethical discrimination that underlay the threatened practices of small-scale mixed farming, suggests the scale of the accomplished reciprocity between inhabitants and their habitat that the flooding of the arable Arrow Valley would compromise. Changes in the moving water were bodily assaults to residents who had learned to reckon direction, duration, distance, and depth by embodying the seasonal movements of the lakes. Commodified water, which moved in time with market demand, was too unpredictable to embody, too quixotic to depend upon as a reference point for the orientation of self in place.

The embodied tacit knowledge and practical ethics that guided small-scale mixed farmers in the rural Canada of the 1950s persist in the Spicer Centre for Sustainable Agriculture and Rural Life, founded in Winlaw in

1999, and continue, as part of the Vancouver-based Sage Foundation, to promote a sustainable Canada.[87] Through their research centre and the Hollyhock Leadership Institute, they continue the work in sustainable, organic agriculture that Jean and Chris Spicer began.

http://
megaprojects.uwo.ca/nuclear

On this site is the new media work associated with the Heavy Water plants at the Bruce Nuclear Facility. There you will find audio compositions that are part of the same series discussed in the introduction to Chapter 3. Whereas those audio compositions dealt with the knowing and teaching of radiation protection protocols, the audio works corresponding to the current chapter, similarly using repetition, overlapping, and elision, explore how residents, cottagers, and plant workers variously interpreted the effects of the arrival of the heavy water and nuclear facilities upon their community and ways of living.[1]

"Baffles" addresses how the nuclear and heavy water plants challenged residents' knowledge structures. Bringing to voice the multiple uncertainties about the activities and dangers at the facilities, the AECB characterizations of various risks, and the residents' and recreational visitors' concerns about the obfuscated inner operations and intentions of the facility authorities, this piece considers how, as one speaker says, "most people were really in the dark."

"Inverhuron Park" assembles points of view on the park as a place of contestation between the rhythms of seasonal recreationists and the inscrutable "whiffs of danger" coming from the Heavy Water Plant.

—JV

6

Uncertainty along a Great Lakes Shoreline: Hydrogen Sulphide and the Production of Heavy Water

Consider the difference between an "eyesore" and a "whiff of danger," both culturally informed judgments about "matter out of place."[2] Place yourself on a street corner on a bright and breezy spring day in a student neighbourhood. The eyesore is the solid and settled line of rubbish at the curbside, the various detritus of a winter's habitation, most conspicuously piles of upholstered furniture befouled by undergraduate immoderation. Then a whiff of something comes to you as you pass into the open space at the end of a lane. Of what it is, you are uncertain, perhaps merely eggs rotting in one of the trash bags. The smell is also like a kitchen familiar, the uncombusted natural gas present in the moment before the pilot light ignites.

My guess is that you would barely pause to consider either of these "matters" your body has registered, and instead proceed home. Body and matter: both are historically specific and susceptible to scrutiny. You know that the eyesore will vanish on a predictable schedule, on that certain spring day when municipal employees come with the right number of trucks to carry "the finite amount of debris." Thanks be to them. As for the smell, it vanishes as soon as you move on, the fleeting scent having offered no certain reason to linger and raise the alarm.

Because the nose is by nature "out there," its olfactory perceptors in a "relatively unprotected position," it readily registers matter borne in the air.[3] Yet smell is a sense that generates uncertainty, more likely to have a phenomenology than a semiotics, to be known directly through bodily sensation than indirectly through symbolic codes like words. Smell presents as an "unpremeditated encounter with the environment and its features." Its messages are sensations held metaphorically, in imperfect analogies, for in European languages smell has relatively few words of its own. Smell registers with the bodily drawing in of air. This "radical interiority" makes the boundaries of the body permeable so that olfactory messages may seem to invade the privatized body, to be particularly intimate, affecting as effect.[4] Smell has a history as warning of contamination linked to practices of self-preservation; its interiority, like that of taste, is historically often a ground for authoritative truth-telling. It can be recalled by place, and yet by nature it is a sensation more strongly perceived when first encountered than after the passing of time.[5] Ironically, these are warning signs that fade even as their source persists. Their qualities and intensity being difficult to hold in memory for retrospective comparison, olfactory meanings are culturally susceptible to being radically remade.[6] Its boundaries being as vaporous as the air that bears it towards, into, and away from sensing bodies, rarely can a smell be placed with precision. As "the sense of transitions, of thresholds and margins," the liminal material qualities of smell have often historically become the stuff of politics.[7]

This chapter is about a place of recreation and pastoral dwelling, low lands below a high ancient shoreline: the eastern shore of Lake Huron at the base of Ontario's Bruce Peninsula. It was used by European settlers from 1852 as a commons, in fall to hunt for migrating birds and to scavenge wood, in summer to find relief at the water's edge from sweaty labours in the fields and gardens above, inland from the cliff. In the 1950s, the Crown gathered up the rights of scattered absentee titleholders to these lands. The federal government built the experimental nuclear generating station at Douglas Point, and the Province of Ontario created a provincial park at Inverhuron. By the 1960s, both the park and the power plant were welcomed as opportunities by local residents and seasonal visitors. The small reactor was a showpiece of leading-edge Canadian technology, presently to be joined by two commercial-scale generating stations as the Bruce Nuclear Power Development (BNPD), a happy source of well-waged work. Inverhuron Provincial Park shared the shoreline and became a holiday favourite for families from all over southwestern Ontario, appreciated for its fine long sand beach, camping places, and well-documented archaeological sites.[8]

In 1969, plans for a new installation at BNPD were announced: a plant to produce the heavy water that served as coolant and moderator in Canadian nuclear reactors. Already, the BNPD buildings and their high intensity site lighting had changed the sensory profile of the shore; the construction work and increased traffic created a certain din. With the heavy water plant came an industrial smellscape. Hydrogen sulphide (H₂S)

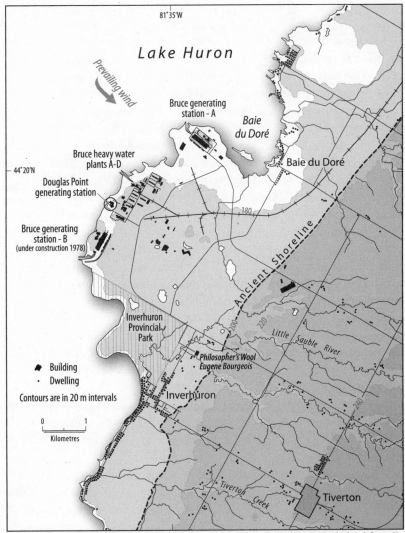

Source: National Topographic Database. NTS Sheet # 41A05 and 41A06 [computer files]. Sherbrooke, Quebec: Natural Resources Canada - Centre for Topographic Information, 2009.

FIGURE 6.1   Map showing the Bruce Nuclear site | Cartography Section, University of Western Ontario, Department of Geography

is an airborne by-product of the process by which ordinary water is made heavy. The human neighbours, campers, and a pastoral family raising sheep and producing yarn adjacent to the site, as well as cottagers immediately south of the provincial park, came to know the heavy water plants by the presence of this olfactory sensation – the rotten egg smell of hydrogen sulphide released from the plants into the air of their living space.

The agency regulating the Canadian nuclear industry, then called the Atomic Energy Control Board (AECB) and later the Canadian Nuclear Safety Commission (CNSC), were charged to license the plants only if, by some combination of empirical measurement and theoretical modelling, their operation could be proven safe. That small releases from the plants might be a nuisance, reducing the pleasures of being in the park and along the shore and of pursing a chosen pastoral life, was not an issue for the AECB. Its statutory obligation was to consider safety, which at that time did not mean avoiding long-term exposures to low levels of H_2S but, rather, ensuring there would not be a catastrophic large release of the toxic waste that would endanger the safe operation of the neighbouring nuclear generating stations and the lives of all nearby. Nor, apparently, was ambient industrial waste a matter of legal resort for anyone at the time since the builders and operators of the heavy water plants were Crown corporations and, by that status, in statute customarily exempt from civil prosecution, including prosecutions under the relatively new Environmental Protection Act (S.O. 197, c 86).[9]

Sensing, science, and statute linked the ways of knowing and reasons for acting, locally, provincially, and nationally, along intersecting paths through time and space. This is a narrative about risk and uncertainty on the permeable boundary between the material and the cultural. Here evanescent olfactory measurement dilemmas and local cultural dispositions clouded assessments of risk, and the historically specific sensing body, as much as policy and technology, figures substantively in an unfolding environmental history.

Three natural elements matter here: the gas, H_2S, which moves through the air with specific physical properties and settles on the ground; the bodies of humans and sheep, which breathe gases; and the site, a place of seasonally variable winds by a great inland sea, backing onto a steeply rising cliff. A conspicuous hybrid in the mix is the Girdler-Sulphide (G-S) process through which the heavy water was made. Among the elements being variously culturally construed were the compatible uses of the site as an industrial and pastoral workplace and as a place of recreation; the aesthetic distinction between tolerable and unsupportable odours; and the changes to be borne locally by all in trade for the well-paying, long-term employment for some in the heavy water plants. The nuclear generating stations do not figure here as contestable, for by the late 1960s they had been accorded a secure place as part of the present and future of the region.

Consider also the two kinds of risk to be appraised: first, the possible release of many hundreds of metric tons of H_2S and its kindred combustion product, sulphur dioxide (SO_2), which might cause many deaths and compromise the safe operation of the nearby nuclear reactors; second, the long-term exposure to low levels of the gas leaking from the plants that could also compromise the health of humans and their animals.

Three sources of uncertainty were present in these interactions: the sensory uncertainty in discerning the provenance and possible hazard of the rotten egg smell; the scientific and engineering uncertainty about whether the AECB models and measurements dependably described the dimensions, dispersion, and density of the plume that would follow a catastrophic release; and the public uncertainty that derived from the statutory secrecy that shielded the operations of both the federal regulator – the AECB – and the Ontario government from public scrutiny.

The Ontario province was the state actor, which, through its ministries and agencies, owned the nuclear site and the park. The human actors in this unfolding narrative included, at the federal level, the AECB and its scientific staff; the legislators of Ontario and the public servants who worked for its electrical utility, Ontario Hydro, and for its ministries of natural resources and the environment; the permanent residents of the region who welcomed work in all parts of the nuclear site; the shepherd whose flocks grazed on its margins; and the long-time seasonal residents in the cottages along the shore, who were captured by the technological sublime of the nuclear reactors but unsettled by the odorous industrial pall of the heavy water plants.

In the story of environmental hazard on the eastern shore of Lake Huron, smell had a starring role. The first "whiff of danger" was sulphur, at first encounter a strong and readily distinguishable odour. But as an airborne industrial by-product, hydrogen sulphide is particularly menacing. It is a hazard "with poor warning properties."[10] At low levels, 0.13 parts per million, it is perceptible to humans as an unpleasant odour like rotten eggs. Ironically, however, at higher and more noxious levels, at 100-150 parts per million, this odour ceases to be perceptible. This is not a matter of cultural habituation but, rather, a daunting visceral dilemma. Oilfield workers learn early in their careers about H_2S, for "sour gas" is a common presence in their trade. At middling concentrations, H_2S kills the olfactory cells, physically extinguishing the sense of smell. An oil worker "knows" he is in grave danger when he ceases to be able to the smell the sulphur, or anything else. The next stages of the physiological and perceptual effects are more dire still. At 500 parts per million, humans experience excitement, headache, dizziness, and staggering followed by unconsciousness (a consequence known in the oilfields as "knock-down"). Respiratory failure follows within five minutes to one hour.[11]

Delayed and chronic effects, particularly of long-term exposure to low levels of H_2S, have been difficult to definitively demonstrate and remain contested, particularly for open air exposures, because ambient levels of the gas are more difficult to monitor there than in contained work places. Because it is heavier than air, when released into the environment at ground level H_2S settles into low areas, persisting invisibly after the release seems to have been borne away by the wind. Survivors of acute exposures have presented with an array of neurological and psychiatric symptoms, including memory loss and depression. The more common intermittent exposures of residents near gas-emitting sites to low and intermediate concentrations (50 to 100 parts per million) yield reports of lingering fatigue, headaches, coughs, hoarseness, and irritability, effects not specific to H_2S and sometimes categorized as subjective.[12] Insidiously, a hydrogen sulphide event is invisible to the eye, perceptible by smell only at low concentrations when its bodily effects are minimal or moot. At high concentrations, an H_2S plume can be either deadly or enigmatically disabling. These material traits made the olfactory perception of H_2S both a source of uncertainty and a ground for suspicion. These were sensations registered in the bodies of the public that, at Inverhuron, by reinforcing apprehensions of secret dealings, eroded trust in public organizations.[13]

In the mid-1950s, the Canadian nuclear project parted company, technologically, with Americans and Europeans with the choice to develop the CANDU (Canada Deuterium Uranium) reactor. These were power-reactors fuelled by natural rather than enriched uranium, and moderated and cooled by heavy water (deuterium oxide) rather than light water.[14] The production of heavy water is one of many contemporary large-scale technologies to which the insights of Ulrich Beck apply. Knowledge of its safe functioning could "be derived only after its construction and operation."[15] The G-S process to be used at BNPD had been developed

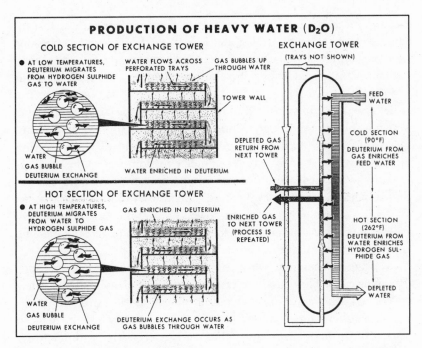

FIGURE 6.2   Flow diagram of GS tower showing production of heavy water | Robert Bothwell, *Nucleus: The History of Atomic Energy of Canada Limited* (Toronto: University of Toronto Press, 1988), 319

and tested at the relatively small military site at Savannah River in Georgia.[16] The Bruce plants, however, would be four times larger than the scale proven in American installations. Their operation would be its proving ground, both the test and demonstration of its safety.

A complex project in industrial chemistry, the G-S process used copious amounts of hydrogen sulphide and fresh water circulating through flow trays in tall enrichment towers. In the presence of gaseous H_2S, liquid water undergoes a spontaneous exchange reaction. At low temperatures, this mixture is at equilibrium when the deuterium concentration in the water is slightly higher than in the gas. At high temperatures the reverse is true. Heavy water is made from light water by a deuterium isotope exchange reaction that exploits this difference.[17]

The Bruce plants were sited near the nuclear stations in order to take advantage of the steam and power they would produce. All were located by the shore because they all depended upon fresh water drawn from the lake. The radical increase in the size of the Bruce plants, by comparison with their American predecessors, proved challenging.[18] In 1979, the plants flared more than 2,000 metric tons of H_2S, and in the early 1980s still flared an annual average of 1,500 metric tons. By the mid-1980s, emissions were successfully abated to about 500 metric tons yearly. Still, from 1987 until the plants closed in 1997, these flare discharge levels persisted at about 200 metric tons yearly.[19]

Because their emissions were blown in the wind, the heavy water plants were not "poster-child" high technologies, precisely situated in time and space, creatures of calculation, exactitudes disciplined by engineering expertise; rather, their hydrogen sulphide emissions were a dreadful uncertainty. While their sources could be pinpointed, their airborne presence eluded specification.[20] Today there are wind farms near the Bruce generating station successfully producing power. During the time of the heavy water plants, these atmospheric characteristics of the site were troublesome rather than constructive.

In spring and summer, when people were most likely to be outdoors, the winds rarely blew from the east so as to carry plant emissions safely out over the lake. The strongest winds in all seasons came from the northwest, over the plants towards the inland townships where in 1979 80 percent of the permanent ninety thousand area residents within a forty-eight kilometre radius lived. The air currents were strong.[21] In these circumstances, tracking the plume of H_2S emissions was a cat-and-mouse game. Jim Dalton, the estimable veteran superintendent in charge of the plant for Lummis, the transnational construction firm that built the facility, observed,

> Given the weather conditions, you had no idea where the plume was going to hit ground. You can make an educated guess, and you do that by sending monitoring teams down the stream of the plume to measure it, to see where it is actually coming down. And so whenever you had a release ... first thing, the monitoring teams were dispatched ... in complete protective gear and they had monitors ... So you could actually plot where the plume was by those results you were getting from the monitoring teams. And that was the only way you had of doing it.[22]

The potentially deadly gas had no colour and made no sound. In this place of strong and shifting air currents, it became one with the wind. Where the plume was, and whether it consisted of merely annoying or potentially deadly concentrations, was vexaciously uncertain.

Experts hoped that smell might tell, but it did not. In the 1990s, the AECB commissioned studies to correlate gas releases with olfactory events.

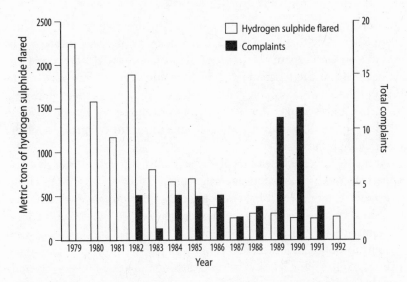

FIGURE 6.3   Amount of hydrogen sulphide flared and number of complaints received, 1979-92. | *Source:* Adapted from original in Michael Prior et al., *Environmental Health Scoping Study at Bruce Heavy Water Plant*, AECB Project 3.168.1. Ottawa: AECB, October 1995.

But through much of the history of the plants, there was relatively little association between the volumes of H_2S flared and the number of complaints from residents who, by smell, had registered the presence of the gas. If winter winds forcefully and unpredictably plumed H_2S away from the plants, in summer becalmed air presented a different potential problem. Then, tropical air masses from the southwest, cooled as they passed over the lake, became stable in their lower layers. Heavier than air, H_2S gas released at ground level could be held in place by the resulting temperature inversions. The most extreme of these, Pasquill stability categories (E and F), present at the Bruce site more than 20 percent of the time from May through September, would confine gas in gullies where creeks flowed through the park into the lake, in the sheltered spots campers favoured, and the low areas of the rough grazing lands to the east where errant sheep might wander.[23]

Ten years after the Bruce heavy water plants began production, uncertainties about the safety of the plants in close proximity to the nuclear generating stations persisted, even among the best informed experts.[24]

Hydrogen sulphide is difficult for the body to know precisely by smell. It also eludes measuring instruments, so that creating a satisfactory theoretical model for the dispersing plume has proven difficult. Troubled by these emerging issues, early in 1969 the AECB convened the Heavy Water Plant Safety Advisory Committee to augment the nuclear competence of AECB staff with experts on toxicity, pollution, industrial safety, and public health from the federal and Ontario public service. The goal was to establish limits for allowable normal releases, define the range of possible accidents, and formulate operating precautions and emergency procedures. All this work was predicated on being able to record the presence of the dispersing plume.[25]

At the AECB, the federal regulator's policy deliberations over this case appear to have given safety precedence. But because its evidence and reasoning were inaccessible to the public, its decisions were often judged arbitrary, captive, or flawed by the citizens whose lives it influenced. Certainly these institutional conventions, as we shall see, led thoughtful citizens to hold fast for decades to ill-founded positions concerning the risk of a catastrophic release.[26] In these arising thickets of misunderstanding, the elusiveness of the plume to precise measurement proved a continuing thorn.

Within weeks, the Safety Advisory Committee concluded that the heavy water plants, their toxic material necessarily separated "from the outside world by only a single membrane," were a greater threat to public safety than the adjacent nuclear station, which had thick walls to contain high pressures and limit radiation leaks.[27] The committee expressed doubt to the power utility that their chosen Bruce location would meet licensing requirements. The designers of the plant were relying on a flaring system for H_2S and its combustion product, sulphur dioxide (SO_2), based on models and calculations of "doubtful validity" for the proposed Bruce gas dispersion system. Two years later, the results of numerous attempts to produce a valid model suggested that a "maximum feasible accident" during adverse weather conditions would result in near-lethal gas concentrations up to nine and a half kilometres downwind. Process leaks of H_2S at ground level, in the presence of the thermal inversions common at the site, would exceed Ontario limits of .03 parts per million at distances as far as three miles (nearly five kilometres) from the source.[28]

If smell did not tell, what about the readings on the gas detectors set out along the most likely path of the plume? Within the committee, members debated the sensitivity of persons to H_2S and whether the reliance on monitors was "as adequate as individuals' own senses," a question that figured in public controversies locally through the history of the site. If doses were being measured at the boundaries of individual bodies, when those bodies had different exposure histories, what would the measurements signify?[29]

The place that posed the most urgent problem was Inverhuron Provincial Park, the 545-acre (220.5-hectare) recreational area immediately south of the rising heavy water plant. Over 200,000 day visitors and 25,000 campers used the park in 1971. Early on, the committee questioned the compatibility of plant and park as near neighbours, and over the following three years it researched the issue with increasing alarm, for the park was heavily wooded and accessible only by narrow twisting tracks. In an emergency, "collecting and shepherding young children who may be swimming, exploring or sleeping in the area" would be a nightmare.[30] While local residents could shelter themselves from the passing gas cloud in their homes or vehicles, campers and visitors would have no timely access to such protection. The park, the Safety Advisory Committee concluded, would require as strenuous protection against a gas emergency as would the plant itself.[31]

The AECB continued the search for better scientific data as well as for a better theory to explain the data, and it commissioned experiments at defence research sites hoping for telling analogies about the possible plume.[32] But fifteen years of lab work and fieldwork left the Safety Advisory Committee with data and interpretations of ambiguous implication, and with a problem that was "untunable." Before a material phenomenon, which kept slipping "through the nets of proof," surrounded by mounting piles of scientific information that underscored their uncertainty, researchers were not confident as to what practical advice to offer.[33]

On 28 July 1971, the Safety Advisory Committee decided to err on the side of caution. The park should be closed during the first season when hydrogen sulphide was introduced into the plant; plant production must not occur in winter; and production must begin only after emergency plans were shown to be fully operational.[34] The Bruce Heavy Water Plant, by then loaded with H_2S, began production in the fall of 1972. Soon thereafter its presence became perceptible by smell to neighbouring residents. They did not respond with one voice. What people smelled, how they interpreted the smell, and what its fleeting presence moved them to do depended on their location both as physical receptors of the sensation and as social beings.

To workers with jobs at the heavy water plant, smell was a positive sign. The plant soon became known in the county as a good place to work, in part because its odour made H_2S "a known hazard," "tangible" by comparison with the imperceptible radiation fields at the nuclear generating stations next door. Workers out and about the plant came to know their adversary. Smell was a warning that, nearby, there might be greater danger that presented no olfactory sensation; smell in a tight spot required immediate retreat. Employees, who all carried seven-minute air packs, worked in pairs, and performed monthly practice drills, came to think of H_2S as a risk they understood and could live with.

There were some memorable moments. Barry Schell, who grew up on a farm nearby, recalled a boyhood day in 1976 when he and his older brother, while hunting groundhogs along the shore, entered an H_2S plume so strong that they rushed to the truck radio expecting to learn of a major event at the site. Later, Schell became a scaffold-builder at the heavy water plant. This was a skilled and creative outdoor trade that gave him great satisfaction. He worked around the flare stacks and exterior piping in the way of both H_2S leaks and plumed SO_2, yet once flare stacks were built, he recalled never smelling the gas again. This may have been a product of physiology (persistent occupational exposures to hydrogen sulphide have been reported to yield permanent olfactory deficits) or of "industrial fatalism" (a form of denial in the absence of good access to reliable information) or both. Schell reported both that his concerns were allayed and, with some irony, that "there wasn't much news given to the general public."[35]

The danger, however, was considerable. Ben Cleary, who came to the Bruce site after working with the gas in eastern petroleum refineries and the earlier Nova Scotia heavy water plants, called hydrogen sulphide "unforgiving ... 400 tons of H_2S circulating around in a unit, if that ever got free ... The Glace Bay plant got shut down in 1969, literally because it wasn't a safe plant." But the many improvements built into the Bruce facility over its history inspired his expert confidence that the employees there worked in safe conditions.[36]

Though with time the emissions were reduced to 10 percent of their initial levels, the smell remained, fleetingly pungent and thus enigmatic.[37] Representatives of the utility refeatured this uncertainty as safety, noting that "rotten egg gas" was a "natural" ingredient in the sour gas processed by petroleum refineries, a smell familiar to high school chemistry lab students, and, like the ammonia in old refrigerators, "as tame as a kitten" so long as it was properly handled.[38] Long-time local residents were disposed to accept these blandishments, for the smell of H_2S was the smell of money. Family security, in the form of jobs, was welcome in a region that was for three generations characterized by outmigration. Like the inhabitants of the California fish-processing centres that Connie Chiang has studied, villagers "learned to turn their noses away." Or, like some residents of nineteenth-century cities, they learned to value airborne wastes as proof of industrial prosperity.[39] Thus the reeve of Bruce Township emphasized the imperative "financial reasons to put up with" the smell. Meanwhile, the newspaper editor in nearby Kincardine deployed his own metaphors to describe the situation: "Now that we've married the project, like a faithful wife who finds out her husband is a cad, we'll have to stick with it and make the best we can. It's too late now to start screaming, the horse is out of the barn."[40] Residents in nearby Tiverton grew unconcerned about the smell and, leaning into officials' most optimistic inflections, distilled the Emergency Measures Organization's (EMO's) warning instructions about the possible range of gas releases into a simple maxim: "they taught us it was a safety thing ... that it's good when you smell it because we're burning off something."[41]

The liminality of the smell made it a physical sensory trace readily refeatured culturally, in this case as an inevitability. This construction of the material traits of the sensuous sign worked in tandem with two other local cultural dispositions towards fatalism. Jim Dalton, the plant superintendent, who as an employee of the construction firm Lummis had experience introducing such facilities into other hinterland regions, noted the first of these: an "ethnic" trait that smoothed the way for public acceptance of the Bruce heavy water plant. The Scots and German settlers of the area, unlike cosmopolitan city protestors, were "very practical people." Eldon Roppel, a Tiverton local historian, made a similar observation: "I think it was like Cape Breton and the coal mines, when you're in a deprived area, you probably have less need to complain."[42]

The second was a disposition invoked by the daunting complexity and unknowable long-term implications of late twentieth-century technologies, a strategic play between being and knowing that had contributed to the earlier decision to locate the first Canadian nuclear power reactor in the Ontario agricultural periphery at Douglas Point, outside the large Atomic Energy Canada Limited (AECL) nuclear reserve at Chalk River.[43] As Dalton observed, "People only want to talk about things they understand. Hard to understand a heavy water plant, even harder a nuclear plant."[44]

For year-round local residents, sensing the smell of rotten egg from the plants became ordinary.[45] Yet such olfactory stoicism was not unambiguously self-protecting. To maintain a keen awareness in the public mind of the inherent liminality of the H_2S smell – that this was a routine by-product of economic development that, on occasion, should prompt extraordinary alarm – was key to the safety of the community. Dalton understood this well. As local residents learned the economic benefits of tolerating a chemical plant in their neighbourhood, many became "comfortable" with the presence of the heavy water plants.

"As you got closer to the plant and Inverhuron," Dalton recalled, "especially in the farming community, there were concerns because anybody spraying manure, specially pig manure, on their farmlands, that would be mistaken for H_2S. So there would be a lot of phone calls coming complaining about the smell. Every one of them had to be investigated ... because you didn't want them to get to the point that every time they smelled something, they thought it was manure, because the two can smell alike at times."

Dalton's safety challenge with the surrounding residents was to retain and work with the inherent ambiguity of the olfactory, to put bounds on the local cultural disposition to acquiesce: "You didn't want them to be comfortable with that smell, even though that was what was happening."[46]

Smell is an ambiguous sensation, vaporous and thus elusive to scientific measurement in the open air, evanescent and thus easily discounted or amplified as a sensuous sign. These sources of uncertainty surrounding H_2S were complicated by contemporary conventions governing public access to information. Until the late 1980s, Canadian governments, both at the federal and provincial levels, used in-house scientific expertise to inform the risk assessments upon which policy decisions were based. This scientific advice was not widely circulated, even within the scholarly community. It was particularly closely held when "essential data to support a preferred policy choice" was lacking. William Leiss, a close student of the problem, has argued that the resultant absence of public scrutiny "was especially useful for covering up the existence of huge uncertainties and the lack of essential data to support a preferred policy choice."[47] Cognizant of the imperfections in both the engineering design of the plant and the scientific knowledge of the dispersing plume, the AECB had done all it could to forestall and constrain the charging of the heavy water plant with H_2S. But the Government of Ontario and its public utility, Ontario Hydro, faced with looming shortages of electricity, succeeded in circumventing the federal regulator and, with doubts about this course of action shielded from public view, hurried the heavy water plant through construction into production.

The shroud of state secrecy hung heavily particularly over changes in public access to the beloved spaces of Inverhuron Park. There was no ambiguity in the terms of the lease approved by the provincial cabinet in November 1972 and signed on 24 July 1973, transferring the park from the Ontario Ministry of Natural Resources to Ontario Hydro. The transfer occurred for one reason: to comply with the safety regulations imposed by the AECB. The ministry would operate the lands as a provincial park for 999 years subject only to limitations specified by the federal regulatory authorities. The priority of recreational use was made plain. The ministry alone had the option to terminate the lease. Land assembly commenced for an additional park at MacGregor Point to accommodate overnight camping upwind from the nuclear site. Recreational day-use at Inverhuron Park would continue.[48] But the terms of the lease and the land assembly were not public.[49]

Two men with whom I spoke three decades later were certain the park could not have continued as a family camping place once the heavy water plant was operating, that the H_2S risk was necessary and sufficient justification for closing the park to camping. Ben Cleary, on the basis of years working with hydrogen sulphide in petroleum refineries and heavy water plants, was one: "It was a big controversy. People were losing that park. But in my own mind ... I think it was better to be safe than sorry." Robert Wilson, a health physicist at the Bruce Nuclear site, and later a senior member of the Health Services Group at Ontario Hydro, was, along with a number of his colleagues, similarly disposed. Between the fission products in the nuclear reactors and the open air were many barriers. At the heavy water plant, there was "no containment ... a much more serious problem."[50] But few among the Ontario public had such insider knowledge.

The Ontario government was involved at many different levels in the remaking of Inverhuron Park. The Ministry of Natural Resources owned the land; the Ministry of the Environment was charged with monitoring airborne industrial wastes. Ontario Hydro owned the nuclear site and had engaged a federal agency, AECL, to build the heavy water plants. The provincial EMO was to implement warning and evacuation plans. All these decisions, among them those about title to the park, rested finally with the provincial cabinet, which, during the oil crisis of the 1970s, was focused on the need for additional nuclear generating capacity. For whatever reasons, these bodies imperfectly shared, both among themselves and with the citizenry, information about the decisions to close the park and the public safety implications of a catastrophic H₂S release. The hope may have been, as science studies scholar Sheila Jasanoff has observed in a similar case of uncertainty, to present "to the outside world" a facade of "quiet authority."[51]

Aside from their accumulating local experience with the odour, the public had some sources of information about hydrogen sulphide. The daily urban newspaper most often then read in the area, the *London Free Press*, carried reports of airborne H₂S releases in Ontario's "chemical valley" at Sarnia, across the river from Port Huron Michigan at the base of the lake, and of evacuations after well ruptures of sour gas in the Alberta oilfields and in the Persian Gulf.[52]

In the summer of 1972, an EMO pamphlet had been delivered to residents within eight kilometres of the site, describing the behaviour of hydrogen sulphide and the low-pitched audible warning signal they should take as a sign to take shelter in closed buildings or vehicles. When in the fall of 1972 six hundred tons of H₂S were loaded into the heavy water plant and production commenced, residents began to hear the periodic testing of these gas warning sirens. Visible signs also clued the public into the potentially dangerous situation. That fall, five new buildings, called "assembly halls" by the ministry and "poison gas shelters" by the press, were installed in the park. Visitors during the summers of 1973, 1974, and 1975 observed that these emergency facilities were staffed around the clock and saw new emergency patrols and more gate attendants about.[53] Cabinet members regretted these signs of grave danger but decided there was no alternative for the period when overnight camping would continue at Inverhuron. The new park at MacGregor Point, up wind from the plant, would not be available for camping until 1976.[54]

Besides the audible and visible signs, there were smells. Mid-way through the winter of 1972-73, start-up incidents made smelling H₂S common south of the plant as cross-country skiers passed the new "assembly-hall-gas-shelters" in the park. In 1973, during the first two weeks of July when the nearby campsites and cottages were full, there were daily complaints of "smellings" along the shore. The exclusion zone that the AECB insisted be implemented around the site, which Ontario Hydro and the AECL preferred to refer to as a "controlled area," restricted further residential development within eight kilometres of the plants, suggesting there might indeed be health issues. Owners began to consider their property values.[55]

Of particular relevance here is that at the highest political levels of the provincial government, smell mattered. James Auld, the minister of the environment, when informed in 1973 that reductions in H$_2$S emissions could not be achieved for another three years, put the chairman of Ontario Hydro and his ministerial colleagues at the Ministry of Energy and at the Ministry of Natural Resources on notice that the current "occurrence of malodours" was unacceptable.[56] But the urgency of the public safety and public relations issue did not prompt full disclosure either to the public or between government agencies. Line staff in Auld's Industrial Wastes Branch, who regarded the amounts of H$_2$S being released as "significant," reminded their superiors in October 1973 that "to date, the public have not been made aware of the frequencies of the H$_2$S losses nor the overall level at which they have occurred," a particularly testy and testing silence given their effect on the users of Inverhuron Park. The supervisor in charge of park planning reported that, "despite close questioning," AECB officials "simply will not and probably cannot commit themselves as to the probabilities" of a "catastrophic release of H$_2$S gas with the wind blowing in the direction of the park during the park operating season." Their colleagues in the provincial Ministry of the Environment could offer only "crude" if troubling analogies between releases of chlorine gas and ammonia.[57]

While ceding little information about the risk assessments, the AECL, the utility's agent responsible for building the plant, and the provincial EMO worked together in a program to reschool inhabitants in the meanings of their bodily perceptions, since these would alert them to signs of risk out there in the open air. Smellings in themselves, officials advised, should not be taken as whiffs of danger. Were they worrisome, the EMO coordinator suggested, these might rather be referred for expert interpretation to the shift supervisor at the police station.[58] These blandishments knitted well into local cultural and economic fatalism to minimize the threat.

Urban campers, coming to vacation in the clean air of the shore, did not perceive the matter in this way. Here were risks with calamitous potentialities and exceedingly small probabilities, risks first knowable, the general public was told in the daily press, by a "rotten egg" odour, which could "cause anything from the simple nuisance of bad smell to unconsciousness and death." The public relations officer for the nuclear site reasserted that the possibility of a serious gas release was "remote," and in January 1973, the executive assistant to the minister of natural resources, surely disingenuously given the November 1972 cabinet decision, insisted that the ministry was "not considering" selling the park. Yet among visitors, the accumulating contradictions gave rise to suspicion. When in July 1973

the senior officer at Inverhuron Provincial Park reassured the press that
none of the releases perceived as "smells" by campers was "serious" enough
to have registered on monitors in the campground, there came a tipping
point.[59]

The facade of quiet authority was crumbling. As the contradictions
grew sharper between what visitors' sensing bodies could apprehend and
what public bodies would disclose, the environmental crisis that the presence of the heavy water plant might bring to campers in the park was
morphing into a "crisis of institutions."[60] In the presence of these apparently conflicting messages, on behalf of the AECL (the federally owned
operator of the plant), Donald S. MacDonald (the respected federal minister of energy, mines and resources) undertook not only to discount the
significance of olfactory sensations but also to question the risk assessment
of the federal regulator that had mandated the emergency measures. In a
letter carried both in the local and the regional urban press, MacDonald
insisted that the AECB Safety Advisory Committee "greatly overstated"
the area over which dangerous concentrations of hydrogen sulphide "can
be expected," and overplayed the "extreme" unlikelihood "that such a
situation would occur." Residents were assured by an Ontario Hydro
technical supervisor that the 0.5 parts per million SO_2 recorded over the
first year of heavy water plant operation would "easily be surpassed most
days" in the industrial cities of the province.[61] MacDonald's intervention
only deepened seasonal residents' senses of suspicion and uncertainty.
Based on Hydro documents, a Canadian Press wire service story out
of Toronto reported two days later that "the measuring equipment on
the site" was not good enough to register hydrogen sulphide emissions
above the legal levels.[62]

Smell was the key to local knowledge of a modern technologically made
event of low probability and high consequence. But between being aware
of a "whiff of danger" and specifying the dimensions of the imminent
threat lay a gap that was eluding the best specialists in the land. For citizens
who were neither chemists nor statisticians, with imperfect information
and a good deal to lose, the processes of inference soon seemed a sign of
bad faith.

Among seasonal residents, Inverhuron Park prompted deeply informed
loyalties. Fritz Knechtel, one of the several bookish offspring of a leading
county furniture manufacturer, had for five decades participated in archaeological investigations of the Archaic and Middle Woodlands sites at
Inverhuron, occupance of which dated from 1500 BC. Summer field schools

run by the Royal Ontario Museum and the University of Toronto through the 1950s and 1960s had yielded fine studies of these sites, forming an exceptional foundation for the popular interpretive program park employees provided each summer.[63] As Knechtel pointed out in his many erudite letters to ministers and the press during the summer and fall after the heavy water plant began production, MacGregor Point was a poor substitute for Inverhuron. If H_2S were really the issue, why was MacGregor Point also well "within the possible smelling area" of the heavy water plant?[64]

Knechtel's appraisal of the risk was different from that of the AECB. Official secrecy kept him from knowing what smell alone did not tell, that the dimensions of the possible catastrophic event were far graver downwind at Inverhuron than upwind at MacGregor Point. And Knechtel was employing a different kind of risk appraisal, using a cultural rather than a technical rationality to weigh the tangible present values of the park against a small and distant inferential danger.[65]

On the long weekend of 24 May 1973, plant managers got unlucky. Canadians celebrate Queen Victoria's birthday on the Monday nearest the twenty-fourth as the first holiday of the summer season. Eighty percent of the campsites at Inverhuron were occupied. Amateur naturalists were out in force, for in late May migrating birds follow a flyway along the Huron shore. Many people would have been in lower boggy areas, where H_2S would settle, seeking out the spring orchids for which the Bruce is renowned. At 2:00 PM on Monday, the accidental tripping of a circuit switch resulted in a release of a quantity of H_2S, overwhelming the capacities of the gas dispersion flares at the plant.[66]

June Ruddock, a bird-watcher, conservationist, and, since 1934, a local cottager, was in Inverhuron Park at that hour with her weekend visitors. For the whole of their two-hour park visit, as she promptly reported to the premier, the "natural environment" of the park was "changed into a chemical environment" by "noxious fumes" from the heavy water plant. It seemed an "anomalous" situation: the province that had been prodigiously spending to improve the park and at the same time urging residents to heat with electricity and buy energy-hungry appliances was now apparently prepared to sacrifice the park to service the engorged energy load it had encouraged. By her account, there had never been "any hint that this odour would occur" and there had been many assurances that the gas shelters had been erected merely "as a precaution." Like Knechtel, she was reasoning with an "expanded vocabulary of risk," facing a state agency whose secrecy and motives she was beginning to suspect.[67] For Ruddock and Knechtel, who were familiar with many Ontario parks, a threat, officially defined as remote, and a mobile and transient sensation known to the public in many places, were being used to justify encroachment on one special shared space.[68]

On the basis of the same evanescence of the sensation and paucity of information that led local inhabitants to make their peace with the plant, the university-educated summer residents concluded that the closure of the park was unnecessary. Partly, this may have been because, from their physical location, they rarely did smell the gas: "It was a quirky, inconsistent kind of pattern ... and of course at that point we didn't even know what we were smelling."[69] Only northeast winds would have carried H_2S over their cottages, and these winds were accompanied by stormy weather that made summer residents happy to be indoors.

Like the people Frank Fisher, Sheila Jasanoff, and Brian Wynne have described, cottagers had grounds to be suspicious of distant decision makers, in this case the Crown corporations, and made their own judgments about the risks at play.[70] Several shoreline residents grew convinced that the closure of the park was merely a way for Ontario Hydro to more thriftily manage its contingent liabilities. Or, more darkly, given the Crown corporation's privileged history as a huge "industrial machine ... run amuck" with "easy access to financing ... so little restraint, so little control" and a vanguardist "engineering management," others put the official silences and fleeting olfactory sensations together and concluded that the provincial government, in the interests of ever-expanding electricity supplies, had authorized a land grab that soon would see the park either paved for a parking lot or cleared as the site for yet another generating station.[71]

In addition to the liminality of the sensuous signs, and the shroud of official secrecy, there was an epistemological gap dividing parties appraising the situation at Inverhuron. The summer residents were mostly humanists and social scientists by training.[72] They were part of the educated postwar generation who had absorbed the postwar state promise of high technologies as products of engineering exactitude and scientific certainty, a promise purveyed by both the AECL and Ontario Hydro. As teachers, the Ruddocks and their neighbours valued and expected clear writing; as graduates of university arts faculties, they may have been unfamiliar with processes of statistical inference and the confidence-levels estimates of risk analysis. For example, Ruddock's husband, William, a professor of French at the University of Toronto's Trinity College, pointed to "the wooly and indeed wily prose of Ontario Hydro and the Atomic Energy Control Board" and demanded not "hasty and shallow" probabilistic contortions but, rather, clear "statements by competent neutral professional engineers." Another cottager demanded the "data" that informed the directives to close the park. In support of the group, Stephen Lewis, leader of the provincial social democratic party (the New Democratic Party), rose in the House to condemn actions taken "allegedly in terms of public safety, although one will never know."[73] Through the summer of 1973, when a series of closely worded inquiries from the staff biologist for June Ruddock's allies in the Federation of Ontario Naturalists forced the Ministry of the Environment to concede that the gas levels within Inverhuron Park were not being measured in the field but determined by "calculatory methods," the suspicions of the cottagers, on their own terms, were confirmed.[74] Thirty years later, the Ruddocks' son, Frank, a Canadian diplomat, remembered with incredulity his parents' Ontario Hydro adversaries: "They knew perfectly well they had an agitated public; they knew perfectly well what they were concerned about; and they were poorly prepared and didn't answer the questions. It could lead to only one of two conclusions: they were prevaricating, or they simply were not all that competent."[75]

Like citizens in many parts of western Europe and North America, the cottagers at Inverhuron were coming to question the imbalance between economic development and the environment that had characterized the postwar technological and political consensus. That summer they did what citizens around the North Atlantic were doing: they organized.

Through July and August, the 250 members of the Inverhuron Committee of Concern circulated pamphlets at the park and in the nearby villages and towns detailing their grounds for opposing the changes in the park. They sold and wore T-shirts proclaiming their cause and conducted a vigorous letter-writing campaign. They mocked Hydro and the AECL as Stalinist drunken thugs, creatures of brutish and mistaken Cold War technological imperatives.

By early August, teenaged Inverhuron Park supporters had collected four thousand signatures on a petition protesting the "Hydro take-over." Though they did not save the park as a campsite, they succeeded in keeping before the local and provincial press their doubts about the utility's evidence and motives for closing the park and their alternative explanation: there was no catastrophic risk, only a strategic opportunity to annex the park into the nuclear site.[76]

FIGURE 6.4   Cartoon drawn by member of Friends of Inverhuron Park, depicting
Ottawa as a fusion of biker and Soviet-era thug | *Kincardine News*, 13 June 1973

By the 1980s, Inverhuron Park, closed for a decade, was gradually slipping "from the public's view and mind"; the regular testing of the emergency warning sirens had become ordinary features of the sensory
environment around Inverhuron.[77] But leaks of H_2S at ground level and
streams of SO_2 from the flawed gas dispersion system, though diminishing,
continued. These chronic trace exposures introduced a new source of
uncertainty for, as Allan Mazur has noted, it remains difficult "to confi-
dently distinguish real hazards of low-level toxic exposures from false
alarms."[78]

Throughout the 1970s, the members of the public most directly affected bodily by the gas emissions from the heavy water plants disentangled "nature at once 'us' and 'other' from us" differently from their neighbours and the park protestors.[79] Eugene Bourgeois and his wife Ann were newcomers, university-educated people who arrived in Bruce County as permanent residents in 1974. They were "back-to-the-landers" who established a business, Philosopher's Wool, hoping to pursue a "traditional, simple, organic" existence, "not quite subsistence but minimalist." Unlike their neighbours, the Bourgeoises' disposition towards the gas threat was not so much comfortable or cynical as cognizant. They settled on grazing lands adjacent to both the Bruce Nuclear Power Development site and Inverhuron Park.

The proximity to the nuclear reactors did not unsettle them. Ann's father, a University of Cambridge physicist who had worked at the National Research Council during the development stages of CANDU, assured his daughter and her husband that "there was really nothing to worry about." Nor did their experience with H_2S in the 1970s start-up phase of the heavy water plant trouble them. Eugene had grown up in Kitchener, a city of meat packers and tire manufacturers. He didn't like the smell of H_2S, but he regarded it as "a nuisance that occurred sporadically ... just one of the by-products of living here." They set about building themselves a safe place, elevating the first floor of their house above the reach of ground-level gas releases.[80] But both ground-level H_2S releases and its combustion product, SO_2, flared from the gas dispersion system of the plant intervened in the lives of the family at Philosopher's Wool. In May 1985, while in his fields, Eugene Bourgeois was overcome by nausea and a blinding headache. He smelled nothing but vividly recalled sensing the gas by a metallic taste in his mouth. In the succeeding months he suffered from symptoms his physicians diagnosed as a central nervous system deficit. Thereafter, the meaning of "smelling the gas" changed for him from a nuisance to "something dangerous," a dreadful indwelling bodily effect. He became fearful of tending his fields and flocks and hired students to help with his work. Twice more, in April 1988 and July 1990, he was overcome in his fields.[81]

The health of his flocks, too, seemed to change. At his farm, through the period from 1985 to 1993, lambs suffered a neonatal mortality rate of between 12 percent and 19.5 percent, compared to the Ontario average of 4 percent, an incidence that placed his flock in the top two to four percentile in a provincewide study. Too many of his lambs were born dead or when born alive failed to nurse and on autopsy were found to have died of malnutrition.[82] A University of Alberta professor of pharmacology who had published on H_2S suggested that the death of the young lambs might have been an olfactory effect, since nursing animals are dependent upon smell to find the teat and apparently hungry lambs at Philosopher's Wool would not seek out a nipple placed directly in front of them.[83]

Only after a 1994 Ontario Ministry of the Environment and Energy study of phytotoxicity in the vegetation on Eugene's land showed that the heavy water plant was responsible for "a minor but measurable increase in sulphur levels" did the family's trouble abate. Plant managers henceforth took care not to flare when the gas might blow over or be trapped by thermal inversions on the Bourgeois land. Workers at the plant, on their own initiative, began to call the Bourgeois farm to warn of impending releases. Thereafter, lambing returned to normal on the farm and the Bourgeois family ceased to smell the gas. Revised 1990 AECB siting guide-lines for heavy water plants, responding to the Inverhuron instance, speci-fied that villages, recreational areas (particularly camping facilities), and topography that would channel a gas release and limit dispersion be avoided when locating future plant expansion.[84] In fact, the heavy water plants had built up such large inventories before their 1997 closing that no further production was required.

Often, then, there is more than meets the eye in environmental histor-ies and histories of technology. Not only will displacing the visual as the primary conduit for environmental knowledge multiply how much we can know about the material and cultural world, but this broader embrace of the sensuous will also offer the opportunity to know differently. In a more sensuous and embodied environmental history, the senses become recog-nizable, qualitatively distinguishable and synergistically companionable, the body both the archive and the instrument tuned to these encounters.

In the story of the Bruce heavy water plants and Inverhuron Park, smell was key, its most developed quality in this instance – uncertainty – a trait of its material form culturally interpreted variously as safety and duplicity. Its evanescent characteristics – in the absence of precise scientific measure-ment and interpretation – amplified doubt, corroded facades of quiet authority, and discounted the fables of political convenience crafted to cover over flawed assessments of risk. This is a narrative in which both smell and the irresolvable epistemological stance "smells like?" were force-fully at work. The visceral and evanescent presence of olfactory sensations was key to how the material and the cultural met. The places where they

met, these technologies, these private and public bodies, and the liveli-
hoods and landscapes they shared were defined spatially and topographic-
ally within a sensorium that privileged smell. The sensorium, the way the
senses are arranged and mutually infuse, here is plainly at once an analytical tool and a character at play in historical events, an element with
particular force out in the open air of the history of technologies and
environments.

http://

megaprojects.uwo.ca/Walkerton

At this site, the new media work associated with the Walkerton water crisis and subsequent Walkerton Inquiry is featured.

While most of the Megaprojects New Media works recounted in this book rely on oral histories and rich documentary materials, the more recent Walkerton case required a different tack owing to the ethical barriers to conducting and broadcasting interviews. The Walkerton Inquiry, which took place between October 2000 and August 2001, provided accounts—on the record—of the various actions and reactions to the recent events. Drawing from the new media work of Jonathan Harris in "We Feel Fine,"[1], the Walkerton project selects, from the 20,000 pages of Walkerton Inquiry transcripts available, certain passages that contain specified keywords. The new media work recontextualizes these by juxtaposing them with other passages from testimony that share keywords or key phrases.

For example, the phrase "good water" in the transcripts ranges from Dr. Pierre Payment's microbiologically founded testimony that "the means of having good water is treatment" to the affirmation of James Kiefer, chair of the Public Utilities Commission, that from one well, "the quality was a lot softer, it was real good water."[2]

In working with keyword queries and a database of text files, this work approaches the task of recollection differently than the other Megaprojects New Media works. However, by using the matching keyword queries to demonstrate a diversity of perspectives among those testifying at the Walkerton Inquiry, this project shares with the other Megaprojects New Media works a suspicion rather than celebration of the reliability of database textual markers as guarantors of meaning, particularly when that meaning concerns tacitly and corporeally held knowledge.

—*JV*

7

Local Water Diversely Known: The *E. Coli* Contamination in Walkerton 2000 and After

In the Ontario town of Walkerton in the week that followed the 24 May holiday weekend in 2000, seven people, all elders and children, died from the effects of *E. coli* 0157 H7 in the town's water supply; in the six months thereafter, half the town's four thousand residents bore the lingering effects of chronic waterborne illness. Blanket news coverage of events in the town was carried nationally through the Canadian media. Both CNN and National Public Radio followed the story in the United States. Soon, far from this one town in rural Ontario's Bruce County, people lost their sense of certainty about good water. As a direct response to the circumstances in Walkerton, but not in Walkerton alone, the political and public disposition and regulatory frame governing drinking water has changed in Canada. Nowhere have these reconsiderations of the ways of knowing and governing good water been as closely followed, or as viscerally and intimately absorbed, as in the town of Walkerton itself, where the tragedy altered local politics, neighbourly relations, and popular understandings of the responsibilities of being and knowing in place. My goal here is to trace how local people, when their search for new understandings was made urgent by sickness and death and made specific by their habits of mutual responsibility in place, relearned what good water was and what the habitus and governance to yield good water must be.

The crisis over water in Walkerton forced townspeople to directly confront directly the clash between internal and external ways of knowing, between distinctions they embodied directly and discriminated among

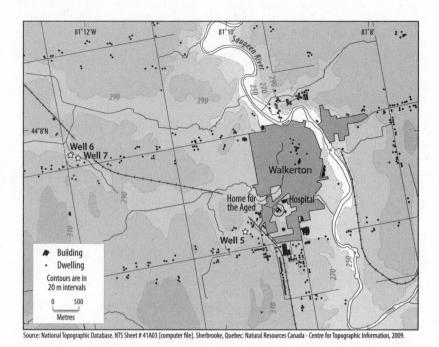

Source: National Topographic Database. NTS Sheet # 41A03 [computer file]. Sherbrooke, Quebec: Natural Resources Canada - Centre for Topographic Information, 2009.

FIGURE 7.1   Walkerton, showing wells and contour lines | Cartography Section, University of Western Ontario, Department of Geography

through their personal histories and local traditions and those of which they became aware indirectly through the mediation of epistemologies presented in words and measurements, symbols and signs conveyed to them by distant experts. We have encountered before this learning to substitute symbolic for somatic authority: in the growing confidence of nuclear workers that the measurements on their instruments were reliable indications of the insensible radiation fields that their bodies did not register, and in the acceptance by parks workers and campers, if not by cottagers, that the sound of sirens around the park at Inverhuron were credible warnings that H_2S had escaped from the heavy water plant and was indeed present along the shore. In both these cases, people were entering spaces that were not their home places. They entered on the condition they obey rules set by authorities they acknowledged: the radiation protection staff in the nuclear generating station and the official regulations governing the parks along the shore. Unlike the learning of new landscapes, streetscapes, fields, woods, and waters required by the displacements from Iroquois, Base Gagetown, and the littoral of the Arrow Lakes, the people of Walkerton, like the cottagers at Inverhuron, continued on in places to which their bodies long had been tuned, where their established traditional, vernacular, community habituation to habitat had served them well. More than at Inverhuron, where the threat to safety could be discounted by multiple uncertainties, complicated by official secrecies, fed by suspicion, in Walkerton the threat to life and health was undeniable. Vulnerable loved ones had died from a poisoning whose effects had touched almost everyone. Not reconsidering ways of knowing self and safety in place was not an option. Practices tested by time had been tragically disproven. Here were environmental and technological challenges to embodied knowledge and local sovereignty that would have to be accepted and accommodated, but how, and to what enduring effect?

I did no oral history in this case. What I report here is based upon the transcripts of the judicial commission convened to inquire into the events in Walkerton and its supporting documents, all now public, and upon the prodigious local and metropolitan newspaper coverage. I attended the inquiry hearings when I could and observed some town meetings. When I undertook this part of the research, the case was sufficiently urgent that sorting out needed to begin even as community feelings remained too raw to ethically tolerate interviewing.

I write about these events both as a scholar and as a daughter of the townships where they occurred. In May 2000, when the *E. coli* 0157 H7 showed up in the town water, I was a guest at a farm near two of the Walkerton wells, 6 and 7, waiting to settle into a field site at the nearby Bruce Nuclear Generating Station. In the 1980s, I had conducted research on masculinity and craft skills in the nearby town of Hanover and thereby was reintroduced as neighbour, friend, and kin to the townships where both sides of my family had settled in the mid-nineteenth century. Since May 2000 I have been in and about Walkerton for several months each year. This study is thus an interdisciplinary intervention, both a conventional historical narrative and, more than the other instances explored in this book, an ethnographic account by a participant-observer. It engages public policy issues as the instance occasioned fundamental rethinkings about expertise and the appropriateness and efficacy of regulation. And because by 2000 I was already deep into this book, my preoccupation with historically specific bodies informed my renewed acquaintance with the town. In 2003, I returned home to take up a position at a nearby university and have continued there as the social scientist attached to the Long-Term Medical Study of the Effects of the Walkerton Water Contamination and as a health geographer teaching and researching about the intersections among technology, culture, and risk.

The first sign of trouble in Walkerton was diarrhoea, a common affliction, often transient, a shared "summer complaint" that would pass without grave consequence before its causes became clear or a more individual implication of kitchen incaution, a short-lived inconvenience non-specifically linked to "something I ate." But the third week of May is still spring, not summer, in mid-western Ontario, and clusters of diarrhoea sufferers were emerging among the children of Mother Teresa Elementary School and the elders at Maple Court Villa (a seniors residence) and Brucelea Haven (the county nursing home). Such a high incidence of illness was not common, and the search for causes and remedies began quickly. The bodies of the young and old are vulnerable to diarrhoea. In the days before the holiday weekend, parents in numbers brought their children to the emergency room of the town hospital. Just in case, the administrator at Brucelea Haven instructed her staff to substitute boiled and bottled water for tap water throughout the facility. Because the manager of the public utility knowingly misled them with reports that the town water was "okay," staff of the local medical officer of health initially focused their search on a food source of the spreading illness. Two days later, on Sunday 21 May, as a precaution, the health unit issued a "boil water" advisory. Only on 23 May, as the pattern of the illness and the Public Utility Commission's manager's deception became clear, was Walkerton's drinking water established as the source of the contaminant that was making people sick. On Friday 26 May, Premier Mike Harris spoke in Walkerton of his sympathy with the townspeople and his "determination to get to the bottom of" the tragedy. That evening, in the home of Veronica and Bruce Davidson, the Concerned Walkerton Citizens was formed. Its first step was to demand from the premier a judicial inquiry into what had gone wrong with Walkerton's water.[3] In the months and years ahead this group became key to the remaking of water knowledge. These people insisted both upon systemic explanations and remedies for the water issues (which claimed national attention through the Walkerton incident) and upon the importance of local voices in the crucial dialogues about risk and belief (which would yield effective water regulations).

Knowing water in Walkerton did not long remain a local matter. All Canadian governments had become leaner in the 1980s and 1990s, and citizens nationally were apprehensive that the tipping point towards too little regulatory oversight had been reached. The ethical conduct of public

FIGURE 7.2   Media scrum outside Sacred Heart Church, May 2000 | Anne Baylin

officials and environmental monitoring of threats to public health were particular concerns. Townspeople soon would be relearning their water from unsettling new perspectives.

The urban media began to file from Walkerton on the weekend after the 24 May holiday. Thus, journalistic narrative conventions immediately began to structure and frame what the public beyond Walkerton would know about the event and what would remain beyond scrutiny. For residents, sorting out what knowledge would safeguard and rebuild the threatened community now became a different, testing, and pressing task.

A reporter on deadline begins quickly to sort out the parties to an event into the roles a satisfying story will require: a hero, a villain, and a mobile chorus whose divided and oscillating opinion will move the narrative along. In the "Walkerton Tragedy," as the *Toronto Star* in its continuing coverage named the event, one of these parties was immediately clear. The hero was Kristen Hallett, the thirty-two-year-old Owen Sound paediatrician who had tested the stools of a boy and a girl referred to her on Thursday 18 May from Walkerton, found *E. coli* bacteria, and immediately reported the results to the Bruce Grey Owen Sound Health Unit. For city journalists, the other two expected players were frustratingly elusive. *E. coli* 0157 H7 was the contaminant that caused the sickness, but the move from cause to blame was disrupted by the contours of the social world in Walkerton.

Stan Koebel, as manager of the Public Utilities Commission (PUC), the putative villain, had gone to ground by 26 May, and testifiers to his villainy were nowhere to be found. The Toronto press made its story through headlines: "People Are Looking to Hang Somebody"; "Water Supply Manager Gets Harsh Words and Sympathy." But the text below the bold type characterized Koebel as an "upstanding member of the community," trusted as a brother by his fellow volunteer firefighters, a decent man whose integrity was beyond reproach.[4] Koebel had been raised in town. His father before him had served as public utilities manager. He met the town's criteria as a guardian of its shared interests. In his twenty-eight years as a civic employee, he'd become "one of the most experienced guys around." Koebel had shown himself capable and hard-working. He took pride in his work, always willing to go the extra mile to make sure things were done. And as his bosses knew by observation, he commanded the respect of his peers among area waterworks managers.[5]

More troublesome for the journalists from away, there were no angry mobs or choruses of blame to be found in town. Without the dramatic momentum of clashing opinion, reporters had to get along with quotes like this one from a Catholic parishioner who would not give her name: "That's not the way here. People want to wait ... There's not time for anger. They have other things on their minds."[6] Those who would speak generally expressed sympathy for a man in the unenviable predicament of having made an honest mistake. Nobody, the media frequently were assured, "meant this to happen." Little changed when Koebel made himself available to meet the press on 30 May, tearful in the company of his family and pastor on the steps of his church. Through his lawyer, he expressed his gratitude for the town's compassion and his trust in the local sense of fairness in the face of "a horrible tragedy." William Trudell, Koebel's Toronto counsel, had read Walkerton well. His characterization of community reaction as "gentle," "understanding," and "non-judgmental" was not far off the mark. The *Walkerton Herald-Times* covered Koebel's return into view only by reprinting the statement his lawyer had prepared.[7]

And so, as often happens with hinterland events, the explanations of what happened in Walkerton fell by default to obliging, if remote, urban commentators. Rolf Helbig of Toronto, writing to the *Toronto Star*, called "the dead of Walkerton ... silent testimony to the irresponsibility and incompetence of rural municipal governments, to the complete lack of caring on the part of much of the local populace and the stupidity of all those who know better but say – and do – absolutely nothing."[8] Rick Salutin invoked the economic historian Karl Polanyi in a Toronto *Globe and Mail* column headed "Walkerton and *The Great Transformation*" to champion the twentieth-century public institutions capable of safe-guarding the public good and to accuse the current provincial government led by Premier Mike Harris of taking Ontario back to the free-markets evils of the mid-nineteenth century. John Gray, in an op-ed page of the Saturday *Globe*, described Walkerton as three hours in driving time and one hundred years in attitude from Toronto, a traditional face-to-face society where outsiders were suspect, local loyalties were strong, and people (close to their rural roots and in their own time) made the best of the hand they were dealt. According to a June poll, 49 percent of Ontarians blamed municipal officials for the Walkerton tragedy. City commentators were inscribing Walkerton as the "rural other," in Gray's words, "the kind of place you think you came from and to which you might some day return."[9]

This characterization comforted urban dwellers. It located Walkerton's predicament in their past rather than in their present. It affirmed the capacity of the modern regulatory state to protect them as citizens. Yet it validated growing provincewide concern that the ranks of the experts monitoring the environment and public health had been perilously depleted.

In Walkerton, the priority was compassionate care. The local paper carried letters, columns, and editorials reminding readers that the town had survived the closing of its principal employer and had saved its threat-ened hospital from provincially mandated cuts. A letter of 14 June affirmed that the "recent disaster [had] opened people's eyes to the true value" of the town's officials. One columnist tried to tame the high-tech news inva-sion with naturalistic analogies: camera lenses as wide as town maples, microphone booms swaying like tall grasses, remassing above the action like flocks of shorebirds. Another asked for news of hobbies and diversions to brighten her section of the editorial page the next week. Most appeals were to not speak ill of the town, or its water, and to tend to the personal neighbourly relationships that made Walkerton special. Concerned to protect those traditions of discretion they would need to rebuild their community on a more sound footing, citizens were wary of premature "finger-pointing" by city media and discomfited by journalists' insistent sidewalk interrogations and their intrusions upon funerals.[10]

FIGURE 7.3   Sign outside the arena publicizing supplies of bleach donated to aid householders decontaminating their homes | Anne Baylin

But already in the local sorting-out of events, there were gestures towards something else. *Walkerton Herald-Times* columnist John Finlay confessed bewilderment that "home" had "suddenly stop[ped] providing the safety and security which are expected." Sue Ann Ellis, reporter/photographer at the paper, wrote of an eerie silence, louder than blades of the medivac helicopters, filling the town, a silence born of uncertainty and disbelief. "We've lost an innocence and we will be changed forever," she observed. "The silence, the tears and the downcast eyes say all those things. Not a word needs to be spoken." Turning a faucet, she wrote, hearing a helicopter, and smelling bleach (the disinfecting agent being distributed for use in high concentrations throughout the town) had become tacit links to an unforeseen and still largely unspeakable consequence.[11] The burden was deathly, the continuing illness painful, pungent, and intimately disabling.

No words needed to be spoken; over the years, much had been said and written. In Walkerton, and commonly in Canadian towns, water was a public issue, its cost, quality, and supply a matter for discussion at election time and at meetings of town council, a topic keenly followed in the local press. Though most citizens had neither the expertise to pass judgements on engineering reports about the state of their wells, mains, and sewers nor first-hand knowledge of the monitoring, testing, and chlorination practices of town public utilities employees, they felt confident, even obliged, to make reports when they perceived the water to be wanting. The terms of the pragmatic agreement about water in this sense were known. And to this extent citizens accepted that they were party to a shared town consensus about water and thus implicated in the consequences if that bargain had been flawed.

Local water is a product of more complex systems than most people in Walkerton and in Canada – as they voted in elections and on levies for municipal improvements, as they maintained their households and drank from their taps – then commonly recognized. In this sense, water bargains were flawed. In Walkerton, as in the contested waterscapes of Ecuador and Spain so tellingly explored by Eric Swyngedouw, the local aquifer was a finely tuned communicator, a receptor of global influences that, by unbidden reciprocity, carried the local out towards the global. Here, un-wittingly, voters and householders engaged a daunting complexity, a co-alescence of mechanical, chemical, human and non-human organic agents, the fusing of social, biological, and technical processes.[12]

For those who gained their knowledge of events through the urban press, what made water differently meaningful in the spring of 2000 was a sense of threat: what was happening in Walkerton could happen any-where. This was the articulation the satellite trucks beamed to audiences far beyond Walkerton. In Walkerton, a more complex understanding that the town bargain about water had been too simple and, by local practice, unsafe was dawning within an altered embodied space. The contaminated water townspeople drank changed the bodily experience of being in Walk-erton. These were days lived amid the merged sensations of pain in the gut and the mingled smells of chlorine and excrement by the many local sickbeds in town; and, overhead, too many times people heard the dark beat of helicopter blades as the most gravely ill were evacuated to the nearest university hospital. In town all these experiences were understood as implications of the long-standing compact the community had made about its water and its guardians. This knowing and sense of shared re-sponsibility, this different articulation, remained local, part of the som-atic and moral compass of the town.

Historically in Walkerton, good water was understood as an arbitrage among taste, softness, and thrift. As Justice Dennis O'Connor, the mem-ber of the Ontario Court of Appeal appointed to investigate the events of spring 2000, noted in his report, the provision of water is a natural

FIGURE 7.4   Helicopter ambulance, the sound of the Walkerton water contamination. Helicopters carried the most acutely ill to hospitals in larger centres | Anne Baylin

monopoly and, thus, unique as a local service. While certain Ontario municipalities in the postwar years had turned their waterworks over to a provincial body, from 1993 called the Ontario Clean Water Agency (OCWA), most towns, Walkerton among them, regarded water as an intrinsically municipal matter.[13] Jim Boulden, a former mayor and public utilities commissioner, told the O'Connor Inquiry: "We had a well run organisation. We didn't need somebody [OCWA] to run it for us."[14] In many places, long after regulating the quality of schooling, medical care, and goods offered in trade was ceded to the competence of distant authorities, good water remained a homely matter, a service nearly free, reliable by custom. Evading the current of centralizing governance, outside large urban centres, the regulation of water frequently remained a closely held community prerogative.

Water tastes. Historically, few people drank water, possibly because from many sources it was unappealing to the senses. But by the 1930s the practices and aesthetics of water shifted. Palatability became an important feature in this newly desirable commodity – drinking water.[15] Aesthetic distinctions are learned. Many people like best what they know best, which is often what they knew first. The contemporary vocabulary of taste is relatively impoverished by comparison with, say, colour. Rather, we commonly recognize and can recall water by place and can be drawn into extended, even acrimonious, discussions of how the municipally treated water of the place where we live compares with water in cities we have visited or with the well or spring water we drink at this cottage or that farm. This too was the case for Stan Koebel and his brother Frank, also a Walkerton public utilities employee, both of whom grew up in town but in a household that took its water first from a spring and then from a well. Into his adulthood, Frank Koebel preferred the raw water to that treated with chlorine, which, by taste, he found less clean, fresh, and familiar.[16]

The appeal of taste as the discriminator of good water is plain enough. Taste was historically the quality by which the value of springs and deep wells had been appraised. Unlike scientific tests hedged by the inferences from a sample collected in some single place in some past time, taste, as a sign of goodness, was an immediate and continuous sign. As they drank, drinkers with each ingestion believed they could judge the water's quality. This reckoning depended upon local knowledge of local waters processed through local bodies according to local standards. It measured safety by the local history of bodily consequences. Most residents discounted the expertise that counselled chlorination, since chlorination compromised taste for an impalpable, inferential, and counter-intuitive promise of safety. That these standards taken from taste and histories of ill effects themselves might be compromised, given that neither *E. coli* nor its associated pollutants necessarily generated off-flavours, and that personal trials with intestinal complaints were, resolutely private matters in town, was a vital connection not made in the chaotic early weeks of the crisis.

In their reasoning about good water, the Koebels were not alone. As O'Connor noted in his report, local residents objected to the use of chlorine and often pressured their waterworks employees to decrease the amount injected into their water.[17] Even as, and perhaps because, the consequences of this preference were unfolding in June 2000 (one of these implications being hyperchlorination to clean the pipes), the reactions of some townspeople against chlorination were strident.

> Has everyone forgotten that chlorine is a poisonous gas that killed, blinded or choked thousands of troops in World War I ... Poisons cause cancer. Period. What really are the long-term effects of drinking, washing and cleaning out homes with chlorine? You can just bet the bleach manufacturers and water treatment plants aren't going to tell us![18]

"Will our water ever be 'safe' again?" Elaine Crilly demanded in the *Herald-Times* of 21 June 2000: "By safe I mean clean and sparkling without chlorination." The Walkerton paper published the same concern the next week, in a letter from Carol Barclay, who, after being sick for five weeks with nausea, headaches, cramps, and diarrhoea wondered, "If over-chlorinated water is the only way to have 'clean water' what then? Chlorine can kill too."[19]

The local aesthetic objections to chlorination were multiple and of long standing. Town officials sympathetically referred complaints about the smell of chlorine from taps, especially first thing in the morning, and the effects of chlorine on clothing to their public utilities staff.[20] These objections continued to be angrily asserted even as daily cable broadcasts of the progressing commission of inquiry made the science of bacterial levels and time of chlorine contact as familiar in town as the plots of the daytime soaps the televised cable broadcasts of the inquiry proceedings had displaced. In early December, soon after Frank Koebel's testimony to his own preference for raw water and his actions to keep the town water palatable on his terms, Brian Glover, one of the commission counsel, was intercepted in a local coffee shop by an elder who queried: "You that fella on TV? ... Well I got a complaint ... there's too much damn chlorine in the water."[21] The long list on the war memorial of town sons who had perished on the toxic battle fields of the First World War and the contemporary experience of chlorine as a corrosive agent made the positive association of chlorine with health and safety implausible.[22]

Historically, the water drawn from wells in Walkerton had been hard, and this raised further aesthetic issues in town. The high mineral content left unsightly iron stains on kitchen and bathroom fixtures and rapidly corroded water tanks. It made bathing and showering less pleasant and doing laundry more of a chore. Chlorination accelerated the precipitation of rust from the water. But the local preferences for softer water and water that bore minimal traces of chlorine were problematically linked. The near source of softer water, Well 5 to the southeast, was shallow, prey to surface contamination, and thus unsafe without treatment. Well 5, as it turned out, was the site of the *E. coli* contamination in May 2000.[23]

These contending aesthetic preferences were complicated by the town's predisposition towards thrift, a trait for which Walkerton was known in Bruce County. Householders in Kincardine, twenty-five kilometres to the west, were charged $57.08 monthly as a flat rate for their water; in Port Elgin, forty kilometres to the north, they were charged $35.20. Residents in both towns paid an additional metred charge for the amount of water they used. Walkerton levied a flat rate. That rate, in October 1999, was $16.50 – half that in Port Elgin, a third that in Kincardine. Keeping costs down for customers was a matter of pride in Walkerton among municipal officials who paid close attention to budgets and expenditures, and who ran their utility conservatively, building up ample reserves. The public utilities commissioners imparted this philosophy to their employees, requiring them to keep staffing and spending as low as possible.[24]

That spring, and in the three years that followed, the people of Walkerton and the wider public who shared in the changing regulatory context of late twentieth-century Canada were learning to think about good water differently, as an unfamiliar complexity. Water shed its definitional clarity. Good water emerged as a something multiply fashioned: by the currents of statecraft; by fiscal decisions about how much public intervention was enough; by private decisions about what water as a commodity was worth; by global definitions of scarcity; and by international judgments about appropriate investments to secure defensible standards of service. Forceful and fearsome within this arbitrage was a new bottom line. Good water must be safe water, and safe water was neither popularly knowable nor publicly certain given the Ontario regulatory frame of 2000. Its security as a life support system could not be assumed without a fundamental reordering of the multiple processes that made water good, a reordering that ceded to public health the priority that custom, convenience, and cost long had naively subverted. The way ahead would be through converging and unsettling acknowledgments that knowledge is not complete, nor its constituting elements necessarily reconcilable, that life must continue amid uncertainty, that neither tradition nor science was a sure refuge in time of trouble.

The residents of Walkerton had not believed that their preferences – for good tasting, sweet-smelling, minimally chlorinated soft water, thriftily acquired through a locally managed public utility – compromised their safety. The employees of the town waterworks regularly drank water from the raw side at the wells; Stan Koebel filled his daughter's swimming pool with town water on the 24 May weekend.[25] And the municipal council was sufficiently attentive to safety that, on learning that the Ontario provincial Progressive Conservative government, led by the fiscally conservative Mike Harris, was about to privatize water-testing services in June 1998, it promptly sent a letter of protest to the legislature at Queen's Park.[26]

Besides taste, and their long-standing experience of the consequences of judging good water by a somatic inference, the residents of Walkerton believed that their well water had been purged of impurities by geological processes. If groundwater, water filtered through an aquifer, was safe, then municipal drinking water drawn from drilled wells was safe. This was a long-standing belief, a verity passed down through generations of water managers in Walkerton – and not in Walkerton alone. Confidence in the safety of groundwater, in the cleansing qualities of the natural overburden around wells, was common throughout Canada in 2000. The beliefs in the reliability of judgments based on taste and the innate purifying properties of drilled wells were sustained by similar epistemological dispositions to reason locally about the direct point of contact between water and the body.

On the basis of vernacular knowledge, the local community had trusted the immediacy of taste to judge good water. Its citizens had long suspected the discrete samples appraised in a distant laboratory as reliable signs of the fitness of the watershed upon which they depended. To incorporate into these appraisals of risk systemic influences such as threats from microbiological contamination or compromised watersheds – conditions that were not palpably and mechanically apparent and needed to be discerned through laboratory results – would have held a spatially specific political consequence. Town and village authorities would have had to defer to the inferential testing regimes of county and provincial experts. Across Canada, outside major metropolitan centres, confidence in local knowledge and concern for local sovereignty made this change seem unnecessary and this consequence unwelcome. Locally validated reasoning, like that in Walkerton, about deep wells and the practices of minimal chlorination, was unexceptional nationally.[27]

Slowly, in the months following that spring weekend, the local unspoken came to speech, and doubt muted technical and policy assurance. Witness John Finlay's bewilderment that home had "suddenly stop[ped] providing the safety and security which are expected"; Carol Barclay's question, "If over-chlorinated water is the only way to have 'clean water' what then?"; and Elaine Crilly's assertion, "You can just bet the bleach manufacturers and water treatment plants aren't going to tell us" about the long-term effects of chlorine.[28] With blanket media coverage of Walkerton, followed closely by news of similar waterborne illness in North Battleford, Saskatchewan, uncertainty about the safety of groundwater became a national issue. Who could know safe water, and what would be the quality of that knowledge from now on?

The authority of custom, the bodily appraisal by sight, taste, and smell that took the risk self-sufficiently in hand, was discredited by illness and death. The apparent alternative, the expertise of science, was neither local nor embodied. Its trade was in the invisible, the impalpable, the only indirectly perceptible.[29] Its method segregated effects rural people suspected might interact. It offered only meagre findings, arcane test results tied to a moment and hedged by probabilistic qualifications. It was stoically silent about the long term and apparently as fallible as a meteorological report in seed-time or harvest. Early in June the *Toronto Star* had reported that water tests were "imprecise scientific instruments" that left local people wondering, "Who can actually conduct a meaningful water test?"[30]

The journalists, lawyers, and scientists brought to town to witness and investigate the growing tragedy had some difficulty assimilating the commonplace assumptions about water in Walkerton. For those from outside the area struggling in good faith to "sort things out," these local beliefs were commonly too alien to be creditable or even recognizable as part of the story. Yet both the lay knowledge of inhabitants long accustomed to observing how the water cycled through the bodies of humans and domesticated animals in the aquifers and drainage basins they shared and the scientific knowledge that guided the interpretation of technological tests depended upon inference. Both kinds of inference yielded results bound by circumstance. For townspeople, the distant experts in the management of news, provincial regulation and public health challenged local knowledge about how humans were implicated in their surroundings. And this expertise, too, it soon became evident, was dissident and contending.

The October Saturday before the O'Connor Inquiry hearings began in Walkerton, the *Toronto Star* published a special report, "The Anatomy of a Preventable Tragedy,"[31] which, by bold headline, accused: "Town Knew Wells Were Unsafe." The headline was wrong. While the *Star*'s eight pages of well-crafted prose and accessible graphics told distant readers much they did not know about the physical and managerial history of the town waterworks, Walkerton residents that weekend learned, apparently definitively, one thing. Because their aquifers consisted of "limestone honeycombed with wide channels," Walkerton groundwater was not safe. By the following Monday, everyone in town was contending with this daunting revelation as the Rotary Club and the Walkerton Community Foundation had arranged for a thousand free copies of the report to be distributed at the three Walkerton convenience stores, the depot at the rink where residents were going for bottled water, the library, and the *Herald-Times* office.[32]

That groundwater was unsafe, if groundwater was unsafe, voided the terms of the town consensus that good water could be judged through taste, softness, and thrift. The *Herald-Times* carried into print the reappraising talk from the street. If groundwater was unsafe, then the public utilities commissioners' careful scrutiny of waterworks expenditures was "nonchalance." If groundwater was unsafe, then public utilities staff measuring out only minimal chlorination were "complacent." If groundwater was unsafe, then local elected officials, rather than having served the town honourably, had "let their responsibility slide in an attempt to save money and win votes."[33] If groundwater was unsafe, then the residents' exposure to waterborne contamination had been long term and the town for years had gone without the protecting local expertise it had assumed. The bedrock of local beliefs about water and risk was crumbling. The residents of Walkerton were aware of their own history. The consensus about good water had been widely shared. People knew that their water rates were low and unmetered and that they had expected their complaints about the taste and smell of chlorine to be heeded. They had supported the search for a source of softer water. It was not easy to construct an urban carapace around the town conscience.

But was groundwater unsafe? The Tuesday before the *Star* special report, Dr. Murray McQuigge, the medical officer of health for Bruce County, Grey County, and Owen Sound convened a town meeting in Walkerton. He summarized the results of the investigative report on the events of May-June prepared by his staff. The historical assumption that groundwater sources were secure when treated only with chlorine needed to be re-evaluated. Groundwater sources in the province were endangered by increasing population density and changes in agriculture and needed to be protected. That afternoon he turned away most questions as premature, citing the impending inquiry. The *Herald-Times* covered the meeting under the headline, "Report Questions Safety of All Deep-Drilled Wells."[34] But Bruce Davidson of the Concerned Walkerton Citizens, the group formed in the spring to press for the systemic answers a judicial investigation might bring, observed that "people were left feeling they'd seen a promo for the inquiry." "I don't think there's any confidence in any level of authority right now."[35] Who was an expert?

Six days after Dr. McQuigge garnered national headlines with his claim that the province's well water system was in trouble, another expert called his assertions "alarmist." Ken Howard, a hydrogeologist from the University of Toronto called to appear before the O'Connor Inquiry on its

FIGURE 7.5   Ron Leavoy of Concerned Walkerton Citizens, addressing a question to
Dr. Murray McQuigge, medical officer of health, in the Community Room at the rink,
15 October 2000 | *Walkerton Times-Herald,* 18 October 2000

opening day, came to a different conclusion from a different scientific
perspective: "There's no reason to suddenly suggest that all groundwater
in the province is a serious problem."[36] In the Saugeen River Valley at
Walkerton, where many people were still sick and the boil water order had
been in effect for six wearying months, disenchantment about the possibility of knowing good water settled in, as unwelcome as a damp fog during
the November ploughing.

Nothing in life is certain, but in Walkerton, and in Ontario, citizens
had come to believe that by supporting prudently run government institutions and paying their taxes, they were lowering the odds against calamity.
If risk were a simple arithmetic of odds, about probability times consequences, then surely the apparatus and the resources of modern public
administration would have made daily life more secure, the way a flu shot
in autumn reduced both the likelihood and the severity of winter illness.
But late modern risk is not about arithmetic. It is a delicate and fraught
exercise in arbitrage. This risk trades our knowledge about the future against
consent to pursue the most desired of our prospects in conditions where
knowledge is uncertain and consent is ordinarily contested.[37]

As the months wore on, and the political, epistemological, and managerial complexity of securing "good water" became increasingly apparent,
the dire consequences of defending local sovereignty became more difficult
to discount.

Many of the witnesses before the inquiry in November were from the
provincial regulators, among these the Ministry of the Environment.

Ministry of the Environment staff members described the document that guided their oversight of waterworks, the *Ontario Drinking Water Objectives*. That water intended for humans should not contain any disease-causing organisms or hazardous concentrations of toxic chemicals or radioactive substances was first on the list of objectives. Detailed specifications quantifying maximum acceptable concentrations for many pollutant types, among them metals, pesticide residues, and microbiological organisms, followed. In mid-paragraph, the document then shifted to aesthetic considerations, acknowledging that "water should be pleasant to drink," and to such quality matters as corrosiveness and excessive soap consumption, recognizing that these constrained the distribution and domestic and industrial use of water. The goal of the *Ontario Drinking Water Objectives* (ODWO) was "to outline the minimum requirements necessary to fulfill" these joint public health, aesthetic, and economic objectives.[38] As questioning from counsel made plain, and as O'Connor found in his report, the ODWO framed as guidelines the criteria that should have been covered by regulations.[39]

The ODWO had been given the muted force of objectives because consent to the most desirable balance among health, aesthetic, and economic concerns for drinking water was contested among the citizens of the province, a circumstance to which the voting record of their elected representatives in the provincial legislature at Queen's Park in Toronto bore witness. Thus did the provincial regulatory regime substantively authorize municipalities to monitor their drinking water as they best saw fit.

At the commission of inquiry local hearings, Ministry of the Environment officials described an incident in Walkerton that made plain why discretionary guidelines and regulations would yield different public health effects from mandatory and enforced objectives and requirements. A May 1998 Ministry of the Environment inspection report had found, and not for the first time, serious problems in the operation of the Walkerton water system: *E. coli* in a significant number of water samples, along with inadequate chlorine residuals, sampling, record-keeping, and training. In her report at that time, the inspector, Michelle Zillinger, emphasized the need for the Walkerton Public Utilities Commission to comply with the ODWO. Stan Koebel promised the commissioners that he would correct the deficient practices. Without question, they accepted his assurances. When the Zillinger inspection document was tabled at the June meeting of municipal council, one councillor, Mary Robinson-Ramsay, suggested that the town's non-compliance with the ODWO would be better remedied by engaging the "regular and on-going technical expertise"[40] of a consulting engineer or a municipal director of public works. She found little sympathy for this proposal, and council took no further action on the Ministry of Environment findings.[41]

Robinson-Ramsay, who taught music and French at the local Catholic high school, was the daughter and granddaughter of physicians who had served the area as medical officers of health. But when pressed in December 2000 by lawyers representing the Ontario government to find fault with her fellow members of council, she would not oblige. Contrasting her own experience accompanying her husband on fieldwork in Africa with the experience of her Walkerton colleagues, she replied, "Maybe they hadn't had a brush with more serious problems," and cast the question more broadly than Walkerton: "Perhaps a lot of Canadians have been lulled into complacency because, when we have had some bacteria, it's been relatively minor." When counsel for the Canadian Environmental Law Association styled the council's stance as "indefinite postponement," she demurred: "I think you are putting it in a negative light. I think that perhaps some of them just really didn't realize the seriousness of it." Given their circumstances, Robinson-Ramsay found credible – even though she did not share it – the risk assessment her colleagues had made, a risk assessment that ranked safety too low among the criteria for judging good water.[42]

Plainly, risk assessments about water in 2000 were being made according to two knowledge systems, one vernacular, dependent on the senses and the local oral traditions by which such sensuous information was passed on, the other professional, dependent on scientific data disseminated in the civil service reporting, which, buffered through the arbitrage of provincial politics, guided discretionary action at the Ministry of the Environment. Robinson-Ramsay had a grasp of how both knowledge systems worked and could serve as an interlocutor between them. Several of her colleagues on the teaching staff of Sacred Heart High School who, with Ron Leavoy, the local printer, and Charlie Bagnato of the town's liquor store, led the Concerned Walkerton Citizens, also showed themselves able to interpret between these different ways of knowing and make useful sense of them for others.

The lawyers at the inquiry were not similarly situated. When it became apparent during background interviews for their appearances at the inquiry that the Koebel brothers and their staff had been falsifying chlorination records, Paul Cavalluzzo, O'Connor's lead counsel in Walkerton, and both the Koebels' own attorneys were confounded. None had experience with fraud that was not grounded in greed.[43] In the inquiry hearing room, many hours were spent reading documents into the record, a process that baffled local residents unfamiliar with lawyerly standards of evidence. Many hours too were spent trying, in vain, to place on record the oral transmission, by time and place, from this individual to that individual, of the unspoken consensus shared in town about how to know good water.[44] The inquiry report noted many instances in which municipal officials had not asked questions about assumptions and actions the inquiry commissioner and his staff thought would have been questionable. O'Connor found that the Koebels had breached the trust of the community.[45]

And yet the Koebels, the council, and the public utilities commissioners had assiduously enacted the local consensus about what was desirable in municipal drinking water. Mary Robinson-Ramsay had told the inquiry as much. They had elected a fiscally conservative member to represent them in the provincial legislature. Through November, it was within this epistemological frame of shared responsibility that Walkerton grieved its lost citizens. Like citizens across Canada and around the world in the 1980s and 1990s, the people of Walkerton had been persuaded by the promises that smaller public sectors would be more effective and private services guided by market incentives more efficient. In Walkerton, these contemporary global ideological currents had amplified rural and small-town confidence that modest self-sufficiency and careful attention to spending were the watch words of good governance. This was the substance of the town's lost innocence. Its consequence framed the next challenge the community faced, a dilemma also reconfiguring the relationships among safety, solidarity, and sovereignty far beyond Walkerton.

That there were two contemporaneous ways by which to know good water, one vernacular and one professional, was not a peculiarity of Walkerton in 2000. In Ontario there were training programs and certification processes designed to ensure uniform, measurable, and controllable standards of knowledge and practice among the stewards of municipal water.[46] But the social compact upholding the integrity of these certification regimes had been weakened by contravening commitments in the 1980s to reduce both provincial spending and central interventions in local administration. To meet these goals, in 1987 the Ministry of the Environment introduced a voluntary grandparenting program that short-circuited the requirement that water operators retrain. Cost-saving and streamlining in intent, grandparenting had the proximate effect of accepting practical experience as commensurable with certification with scientific and technical training.

Thereafter, there was no enforced "common set of practices" binding operators in a shared monitoring culture.[47] The Koebels' certifications were regularly upgraded thereafter as the mandated requirements to run a system such as that in Walkerton were raised. In the years that followed, as Justice O'Connor noted in his report, the "MOE took no steps to inform them of the requirements for continuous monitoring or to require training which would have addressed that issue."[48] Through the 1990s the provincial government, focused on disentangling provincial-municipal relations and trimming its regulatory apparatus, paid little attention to water.[49]

The Koebels, and the many other waterworks operators with grandparented certifications, were not running their systems with "unguided practical experience," as some have accused. They were guided by experience their communities shared and respected. They acted not in the absence of knowledge, in a vacuum calling out to be filled, but on a series of propositions that were held in common and that were integral to the valued cognitive sovereignty their community.[50] These propositions were part of the connecting tissue that held the town together. Shearing them away would be consequential.

Near Well 5 was the farm of the town's veterinarian, Dr. David Biesenthal, and his wife Carolyn. The Biesenthals kept a small herd, and manure from their herd contained bacteria that matched the *E. coli* 0157 H7 found in Well 5. In the week of 8 October 2000 the town had celebrated the return to Walkerton of their daughter, Laryssa, a bronze medalist at the Sydney Olympics. That same week, Stan Koebel and the PUC named the Biesenthals as a third party in a $350 million civil suit.[51]

In early December, as Frank Koebel's history of habitually falsifying chlorination reports entered the inquiry record and the town re-weighed both the burden of its lost innocence and the question of blame, a number of agricultural groups began to solicit support for a trust fund to assist the Biesenthals with their legal bills. As herders, the Biesenthals were model practitioners of modern scientific risk abatement strategies. As the *Herald-Times* reported, quoting a press release from the Bruce County Cattlemen's Association, the veterinarian and his wife had "taken numerous steps to ensure that their farming practices" were "as safe as possible" and had been among the first farmers in Ontario to complete an environmental farm plan.[52] To blame the Biesenthals, as if they "had played an active role in what happened,"[53] seemed equivalent to blaming every person in town who, when ill in May and June, might have unwittingly contaminated their neighbours and kin. Surely, John Finlay affirmed in a

6 December column entitled "Compassion Goes Missing When Money on the Line," no single person or event was going to emerge as the villain in the Walkerton story of 24 May 2000.[54]

Following the money became a preoccupation in town, the mundane materiality of this pursuit a clear contrast and complement to the fugitive, if now speakable, apportioning of blame. On Friday 8 December 2000, Walkerton learned for certain that Stan Koebel had negotiated a $98,000 settlement with the PUC and, by implication, that his PUC salary had been $69,000 per year. In a local economy where many earned little more than the minimum wage, where Betty Borth, whose career as an operating room technician at the hospital paralleled Koebel's at the PUC, had been paid less than half Koebel's salary, the sum startled. The following Monday, the Concerned Walkerton Citizens convened a meeting to frame a petition calling for a retrospective health study of the effects of the town's incontrovertible history of contaminated water. By force of circumstances the meeting turned to re-reckoning the consensus about good water. While there could be no certainty about whether residents' histories of bowel problems were linked to the water, as stories of cramping and diarrhoea endured covertly over decades came to voice in the hall, strong and certain opposition formed to Stan Koebel's severance settlement.[55]

Although his appearance was delayed by deliberations over his mental state, Stan Koebel, sedated and tearful, testified for three days beginning 18 December, setting into the inquiry record a detailed account of his troubling stewardship of the town's water. The urban press responded with outrage. A *Globe and Mail* editorial refused Koebel the sanctuary of ignorance and condemned his behaviour as a "sort of calculated deceit." The *Toronto Star* published a stinging parody by Joey Slinger of a small town where nepotism, drunken debauchery, petty violence, and fractious complicity prevailed along streets piled high with manure. The *Star* also published an editorial cartoon twinning Koebel with Homer Simpson under the heading "Separated at Birth?" But both city papers acknowledged that rural idiocy was not a sufficient explanation for events in Walkerton, that citizens of the province were separated from a similar catastrophe only by a few overworked civil servants and a misguided government besotted with thrift – an unhopeful appraisal of the efficacy of late modern risk management in Ontario.[56]

In Walkerton, outrage was more of a luxury. As with any luxury, there were those who found themselves in a position to indulge and who did so, at least for a time. Charlie Bagnato, the manager of the liquor store,

SEPARATED AT BIRTH?

FIGURE 7.6   Cartoon of Stan Koebel. *Toronto Star* cartoonist Theo Moudakis gives a new twist to the common assumption among urban people in 2000 that events in Walkerton were a product of rural idiocy | Theo Moudakis, *Toronto Star,* 19 Dec 2000, A24

and a man who knew himself to be excitable, found the outraged passing through his doors, "the usually conservative Walkertonians, very quiet," and he in the unfamiliar position of "having to calm them down." Walkerton, being Walkerton, there were many who thought about luxuriating and then re-thought, and there were some for whom the thought was unthinkable. Ron Leavoy, the printer who headed the Concerned Walkerton Citizens, was one of these. Leavoy thought Koebel, whom he counted a friend, "was being scapegoated when he was far from acting alone in his role in the tainted water." "I don't think there was any malicious intent there at all. Most people in town are that way too. They know he made a mistake, but he didn't intentionally go out to harm anybody."[57]

John Finlay, the *Herald-Times* columnist, by avocation a member of a Christian Peacemakers team, put in print the reasoning of many in Walkerton. Frank Koebel believed in the town's water, as many others in Walkerton and in other communities had for 150 years: "This faith was either misplaced or a bit too old-fashioned. It didn't keep pace with changing technology, land use practices and other environmental considerations." But as the people of Walkerton had shared in the consensus that good water was defined by taste, softness, and thrift, so they shared responsibility for the tragedy of 2000, "both those who betrayed their trust through their actions as well as those who did so through their acceptance of the way things have been for a long time."[58]

Finlay did not mention Frank's brother Stan in this summation. Anger towards him remained keen, his severance package a continuing sign of trust betrayed. The municipal council of Brockton, a new entity formed by provincial streamlining initiatives to merge the town with surrounding rural townships, dissolved the Walkerton PUC in January 2001, thereby hoping to prevent payment of the terms of termination to which the PUC had agreed. In their final severance settlements, Stan received $48,000 and Frank, $55,000. Management of the town's waterworks was turned over to the provincially run Ontario Clean Water Agency. Physicians from the University of Western Ontario Medical School undertook a seven-year study of the long-term health effects of *E. coli*. And the government of the province has committed $50 million to support the Clean Water Centre of Excellence to be located in Walkerton. In April 2003, the Koebel brothers were charged with reckless endangerment, breach of trust, and forgery. But the lessons learned in Walkerton are spreading slowly. December 2002 reports of the Ontario Ministry of the Environment show that half of the province's water treatment plants are still in default of proper testing protocols and in violation of the safety rules implemented after Walkerton.

And so the story ends and does not end. Many residents of Walkerton still will not drink water from the municipal system; some vow they never will. Elements of the local tradition about how to know a good man have remained. Stan Koebel has moved away from Walkerton, but Frank has "stayed in town to face people and people respect him for that."[59] Troubled by the uncertainty inherent in the late modern assessment of risk, by the authority they have ceded to distant regulators whose methods and necessary compromises they will not know, townspeople live on in a present where the preconditions for their safety are impalpable and, thereby, a matter of scientific inference, chastened by their own histories of trusting consent. A widely supported "Healthy Communities" initiative is at work, laying new foundations for solidarity in town. Members of the Concerned Walkerton Citizens have travelled widely to share what they have learned with First Nations band leaders and water advocates across Canada. More intimately than the city journalists and their readers, or the lawyers and social scientists who so effectively served the O'Connor inquiry, more viscerally than most citizens in the early twenty-first century, the residents of Walkerton know the doubt at the heart of a society organized around risk. And to their credit, they have acted in conscience and have not pushed that troubling knowledge away.

In Walkerton, taste was an authoritative vernacular knowledge, trusted as a guide to direct, embodied, and unmediated truth about water. Such knowledge is made and remade within the social and cultural history of the sensing body and was being remade in Walkerton as the events of 24 May 2000 discredited taste as a test for safety. To know good water was no longer a privilege of commitment to place, a feature of a body adapted to discern local truths. The new definition of good water as safe water depended on scientific expertise tied more to profession than to place. Thus, the derogation of the authority of taste was a challenge to local confidence in certain local ways, to this locally embodied knowledge, and to local sovereignty. Water bore the current that reshaped understandings of safety and sovereignty in Walkerton. Like a tide, it flowed two ways. Water carried far from town the implications of local beliefs and practices. These, in turn, were being remade, both before and after the tragedy of 24 May 2000, by international influences. Good water emerged as a complex of decisions about risk guided by knowledge and hedged by uncertainty.

8

CONCLUSION
Historically Specific Bodies

This book is about bodies adapted to time, place, and practice. First borrowing insights from sensuous anthropology and its progeny, sensuous history, which take the five senses as specific distinctive conduits between humans and the "worlds beyond their skin," and then building upon advances in cognitive psychology that understand perception as contextually tuned, as functioning synaesthetically through interacting networks (rather than as "hard wired" into distinctive channels), the goal is to put these conceptual possibilities to use. The move here is from the realm of theory to the materiality of the everyday. The cognitive and cultural reordering entailed in adaptation to changes in natural and built environments, technologies, and the patterns of daily life was both barely conscious and momentous. Inhabitants would have been required to do more than alter the repertoire of sensations they recognized and to which they gave meaning. The synthesizing capacities of their bodies would also have to rehone their sensibilities, remaking the integrating modes of attention and perception through which they made sense of their circumstances. The deep internal disturbances people weathered, as their historically specific sensing bodies retrofitted, would have been difficult to account for, registering first as bewildering errors, lapses in mature judgment, and child-like failings to pay sufficient attention, to plan prudently. These closely textured studies consider these individuals and communities (many of whom became allies, co-investigators, and animators in the research) and follow the transformations in their bodily archives of knowledge, habit, and reflex as they were tested, replaced, renovated, and renewed by practical encounters with their altered habitat. The goal is to bring out the importance of locals' bodies and senses in orienting them to these places they knew as well as to the new, altered places that emerged in their midst.

Making researchable these relationships among sensuous understandings, embodied legacies, and the accommodations to the exercise of disembodied power was possible only by sharing authority through collaborations with those who were there before, during, and after the displacement and dispossession. This embodied knowledge, awarenesses held by those who weathered the challenges of the new and knew the retuning viscerally, often emerged by reading the documents against the grain and by listening for what was lost in transcription. It has meant supplementing customary historical analysis of transcripts, images, and landscapes with analytical devices that can register the experiential excess arising from these radical declensions, the gathering up of the gestures and inflections that have resisted the rhetorical razor and been pushed to the margins by the conversion to text.

Happily these instances demonstrate that the reservoirs of revisable tacit knowledge that Harry Collins and his colleagues have successfully studied among experts in laboratories are also recoverable by and from those who might think themselves ordinary folk. Like the white-coated bench workers of Collins' accounts, and the mechanics in overalls whom Douglas Harper got to know as they repaired Saab automobiles in upstate New York, the people who have been the focus of this book, perhaps particularly because their attention had been concentrated by a radical reordering of their everyday routines, could recognize how their bodies had been tuned to time, practice, and place, and identify the dissonance the new order made between them and the world beyond their skins. Embodied approaches can lead researchers to fuller understandings of the burdens that rapid technological and environmental change place on "people in the way."[1]

This path towards knowing anew is difficult to share, for these different embodiments of the altered world beyond our skin are usually achieved without conscious awareness and are held beyond speech. When changes are momentous, as in the cases considered here, the retuning of inferences from perception, the remaking of visceral muscular responses, and the reforging of practices for everyday life can be testing. Some emerge from the process relatively soon, radiating renewed energy, with a spring in their step and projects to plan. For others, these are traumatic burdens, crises of competence and confidence carried for decades as personal failings, presenting only as retreat into private sorrows and away from community institutions. After calamitous events, these most interior sources of sadness can be slowest to heal, difficult even for intimates to access, and profoundly elusive to researchers who are always in some sense strangers from away. Some of these challenges are more accessible through our new media experiments[2] than through text – the air photo series registering landscape change at Gagetown, the sound essays suggesting how residents process their revised awareness of safety and danger, the walks around Iroquois, the making and remaking of Val Morton's ranch south of Nakusp, the stylized Walkerton Inquiry excerpts illustrating different notions of good water. Others become clearer, by inference, from the collective strategies the "people in the way" developed to assuage their losses and establish themselves anew.

In each of the places, examined here as cases, people in their own time found resources for resilience. The pace varied with the circumstances of the unbidden upheaval that forced the remaking. The paths they followed were distinctive from one another, enabled by specific local means and differential access to more distant help. These contextual elements of response and repair, external to the processes of embodiment, are accessible and researchable. They informed the futures of these places; they are what we have. To examine them thoroughly would be another book but, *pour encourager les autres*, I offer a beginning here.

Surely the matter of consent was a key to how "people in the way" weathered challenges to their embodied balance in place. Yet consent was a temporally specific constraint leveraged through the duties of citizenship. The reciprocal promises of a shared national inheritance were bargains that would not and could not withstand the test of changing politics and interest. The megaprojects of Gagetown, the St. Lawrence Seaway, CANDU, and the Columbia Treaty dams were birthed in the urgencies of the Cold War and the aspirations for postwar renewal. These spawned politically mandated commitments to North Atlantic defence, a national welfare state, accommodation of New Canadians, and the reduction of regional economic disparities. The sacrifices required in pursuit of these goals were unequally distributed. Like urban dwellers who found themselves in the way of expressways, or happily resident in areas others specified for renewal, these rural and village inhabitants found their own needs divergent from the priorities of central planners at a time when economists and engineers, guided by professional commitments to the "highest best use" of resources, made local needs axiomatically subsidiary to national goals. Those in the way who posited alternatives – a training base on the unsettled lands at Utopia in New Brunswick, Wells Coates' plan for a new Iroquois close to the St. Lawrence along the heights at Flagg Creek, a Low Arrow dam and a smaller reservoir to protect established settlements and land use along the Columbia, continued access to Inverhuron Park – found that development plans left no conceptual space for more than one way. Promises made – that the dams and the simulacra heritage sites assembled using structures salvaged before the rising waters would attract tourists, that industry would locate near the new hydroelectric and nuclear sites – were early suspected and in time proved flawed. This was not an era of open government. The megaprojects could not have been effected at the pace or on the scale national military, economic development, and redistributive priorities then denominated if fully informed consent from the citizenry had been required. As experts, some public servants caught in the slipstreams of the development goals – Robert Wilson at Ontario Hydro, Jim Wilson at BC Hydro, Norman Wilson of the Second World War advisory committee on the prospects of the Seaway, and A.K. Das-Gupta and his colleagues called to serve on the Heavy Water Plant Safety Advisory Committee – knew more both of what was likely to follow upon the development of the megaprojects and what it was impossible to know with scientific certainty. Yet they could not budge their superiors from their stated intentions. If citizens agreed to or deferred to the massive engineering works that displaced them or were set down in their midst, they did so with imperfect knowledge and amid grave uncertainties.

Among the instances examined here, I suspect that the challenges to residents' senses of competence and confidence were most acute in the lands back from Gagetown, and along the St. Lawrence and the Columbia, where residents' embodied reference points had been taken away. At the nuclear plants and in Walkerton, there were teams of experts to shepherd the process of knowing anew. More commonly, in the era of megaprojects, no one really knew. There were inklings of what the future might bring. Think of ninety-three-year-old Elizabeth McCarthy asking from her home along the Lawfield Road, "What price do they pay for contentment?"; Cooke Sisty's father observing in awe and dismay, "Son, they're changing the face of the earth"; Robert Roder's premonition of "much tragedy and bereavement and heartache" following from the drive along the Columbia "to commercialize and make profit out of every good gift God has given for the benefit of all." But the inklings did not capture the intimacy of the impending trauma, the undoing of what was bred in the bone, the unease Joyce Fader still felt long after the inundation: "When the land is gone, then ... you can't go back." It is surely no mystery why most took a decade or two to regroup. Yet regroup they did. Though some died, and some irretrievably lost their bearings, many, like Joyce Fader, in the absence of alternatives simply held on: "You just have to keep going ahead, regardless of what happens in your mind." By "going on" they variously summoned resources for resilience.

Many had been just where they were for decades, some for generations. Though they had lost the taskscapes and landscapes they knew viscerally – if, in the case of Walkerton, mistakenly – they retained advantages from shared long settlement. They knew what human resources they had to work with. Certainly these resources were diminished by recent events, the out-migration, the illnesses and deaths, the losses of familiar lands and waters. But together most had previously weathered good times and bad. There were good jobs for the current and the next generation at the nuclear plants, and the same state that had undone residents in several of these sites offered mitigations, some curious and ill-fitting but possibilities nonetheless.

At Base Gagetown, there were steady, well-waged jobs for support staff and, for a decade, seasonal work for men who knew the pastures and the woodlots. Their skills were valued when it came to building the training grounds, by mechanical means clearing the mazes to test the skills of novice tank operators and opening the ranges where the accuracy of the infantrymen deploying weapons could be honed. When the Canadian and American departments of defence turned to chemical defoliants, among them various dioxins, the implications of this promised employment became more complex and, as we have seen, are still unfolding.

As the years passed, except for the cemetery visits on the military Remembrance Day, 11 November, the base lands were closed to former residents. The sounds of explosives and the rumbling of tanks proceeding up access roads into the base were signs of some radical remaking of the territory left behind beyond the wire. Locals who worked on the base and military personnel shared news of changes. Regional newspapers, in annual stories of massive manoeuvres, included some photos in which dwellings, churches, and landforms were identifiable and others in which, ominously, nothing that was familiar remained. Outside the base, the fiddling and dancing continued, sometimes in familiar halls, such as the Dunn's Corner No. 4 LOL, relocated along the river at Queenstown. There were ballads memorializing the "exile" from the base lands amid the elaborations of old reels and new compositions.[3] But by 1983, some more systematic reclamation of what had been lost became possible and necessary. Individuals had begun to write their personal and family histories, and many gathered in 1983 for a thirtieth anniversary reunion. In July 1998, the Base Gagetown Community History Association was formed, and, since then, histories of the twenty communities of the baselands have been published, many completed especially for circulation at the four-day event to mark the fiftieth anniversary of the expropriation in 2003. There was federal and provincial support for oral history projects about social, family, and institutional life in the lands that became the base. The association maintains a website that guides readers to genealogical resources, family histories, newspaper clippings, photographs, and paintings of the baselands as they were before the base came, with open invitations to visitors who wish to add to the archive with research of their own. All this activity, and the continuing dances and musical evenings, make vital and present what fifty years ago seemed the lost world beyond the posted boundaries around the base that replaced the fields, meadows, and woodlots of the former parishes of Gagetown, Hampstead, and Petersville.

At the CANDU sites, the challenges of change have been easier to accommodate, at least in the first fifty years. The lands by the shore, below the ancient lake edge at Inverhuron and on the Point at Lepreau, had not been intensively used before the generating stations were built. Some lost access to traditional common space where they had fished and cut firewood, but the promises of well-paid, skilled, and long-term employment have been realized. At Bruce, many men and women who graduated from local high schools in the 1960s took jobs as service maintainers. Some rose through the ranks, through courses first at Chalk River and later at a facility on the local site, and learned to know new workplace dangers through in-service training. The next generation went off to the University of Waterloo, the University of Toronto, or McMaster University for degrees in geophysics and nuclear engineering or to Fanshawe College in London for courses in nuclear technology. Rather than moving away to the cities, as their forbears had done since the late nineteenth century, they returned to their home places and to rewarding, challenging careers at the generating stations. The area prospered, though the local economy became less diversified as long established manufacturing concerns lost skilled workers to the nuclear site, though the new horticultural and feed processing firms, drawn to locate near the plant by the expectations of concessionary power rates, foundered when these hopes proved ungrounded. The spent fuel rods from the reactors, accumulating in pools on the Bruce site since 1968, are commonly viewed among area residents less as a threat than as a responsibility reasonably borne locally in exchange for the benefits of nuclear employment. Municipal politicians from the area, led by the mayor of Kincardine, have spearheaded a national organization of nuclear communities, insisting that, in contrast with the 1950s planning to site the reactors, decision making about the disposal of all categories of radioactive waste will be consultative. Leaders of the nearby First Nations, the Chippewa of Nawash, guided by the chief of Band 27 at Saugeen, a lawyer with experience in a leading national firm specializing in Aboriginal law, have successfully negotiated an agreement with Bruce Power (the private firm that succeeded Ontario Hydro as proprietor) to recognize aboriginal stewardship and entitlements over the lands and waters around the Bruce Peninsula.

The Seaway, by contrast, cut through sites and drowned structures both intensively used and key to the rewriting of the postwar national narrative. Local people were promised jobs in the series of parks along the St. Lawrence between Kingston and the Québec border (touted as a new "front door to Canada") and in the Upper Canada Village heritage site between Morrisburg and Cornwall. The national purpose was to "memorialise a great people – the United Empire Loyalists – and a great battle – November 13 when valiant British and Canadian soldiers scored a decisive victory in the Battle of Cryslers' Farm," through the restoration, "to an original but not modern lustre,"[4] of Georgian mansions, a Protestant church, and a grist mill. Historical consultants were unsettled by the plan to create "a sugar coated pill," "weakest in the field of things intellectual." For them the drive to locally absorb "surplus manpower" pushed the pace of the reconstruction of the more than forty buildings that were removed from the site ahead of research, and "the preparation of detailed working plans"[5] created an inaccurate folly rather than a usable guide to the past of the place.

The issues that troubled area residents dislocated by the Seaway were grave sins of omission, personal rather than professional affronts. By the mid-twentieth century, most of the descendants of the eighteenth-century exiles from the American War of Independence had moved on from the region to the urban centres and better agricultural lands farther west.

Textile mills then located along the St. Lawrence. These drew experienced
French Canadian workers and their families from the Montreal region and
the eastern townships of Quebec. Only the Caldwell Linen Mill at Iroquois
and the nylon plant at Cornwall weathered the inundation. These people
were both dislocated by the decline of textile employment in the new
Seaway Valley and excluded from the publicly authorized repositioning
of the region's history.

Fran Laflamme, daughter of the barber in the drowned village of Wales,
settled after the inundation in the merged new town of Ingleside and, as
a teacher-librarian, quickly became active in a number of community
building pursuits. In 1977, she drew together others displaced from the
villages of Mille Roches, Moulinette, Wales, Dickinson's Landing, Farran's
Point, and Aultsville; the hamlets of Maple Grove, Santa Cruz, and Woodlands; and the farming community of Sheik's/Sheek's Island. They called
the association they founded the Lost Villages Historical Society. And on
their bilingual website, they forcefully asserted that their villages were "not
lost through carelessness" [but] "disposed of with Government approval
'for the common good.'"

To memorialize the nineteenth-century inhabitants of the Seaway site,
the association built an alternative village site to Upper Canada Village
at Ault Park, on the new riverfront back from the drowned settlements,
and named the road up to the site after Fran Laflamme.[6] That the next
president of the Lost Villages Historical Association, Jim Brownell, has
since been elected as representative of the area to the provincial Legislative
Assembly is some indication of the broad support this popular alternative
to Upper Canada Village shares among local voters. The area between
Cardinal and Cornwall still looks to travellers more like a map of a place
than a place itself as the soil of the district is shallow and in many areas
overlain by sterile construction debris. While the physical and social
burdens of the displacement linger, the next generation is stoutly forging
a place for itself back from the interesting river retrofitted as a channel
for shipping, perhaps more affirmed than abashed by the fact that the
Seaway – the megaproject that residents opposed fifty years ago – has
outlived its economic usefulness.

In the Columbia Basin, of which the Arrow Lakes are part, fifty years
on, the state partners to the Columbia River Treaty (CRT) have explicitly recognized that the agreement was signed without prior consultation
with the people of the Basin and that "most of the benefits of the CRT
were enjoyed by areas outside the Basin while most of the negative effects
were, and still are, felt by the Basin and its residents." Upon pressure from

local residents,[7] the 1995 CRT included an instrument, the Columbia Basin Trust, managed by a board of Basin residents to redistribute locally "a fair share of the ongoing downstream benefits earned" under the treaty, $250 million as an initial endowment and an additional $2 million per year from 1996 to 2012 to be spent on "social, economic and environmental benefits for the residents of the Basin." The work of the trust has included projects "to conserve and enhance fish and wildlife populations affected by construction of BC Hydro dams" through habitat conservation and restoration, and funds for regional economic and cultural development.[8] The modern village at Fauquier, on the east side of the narrows between the Arrow Lakes, which the displaced families judged odd and unsuited to their ways, has been populated by migrant retirees drawn by the golf course. The bridge across the narrows remains an unfulfilled promise. Current disputes about "highest best use" concern the future of the forestry.

Inverhuron Park, closed to camping for twenty-nine years, reopened 125 campsites in July 2005. Ontarians who were promised, "if you remember camping at Inverhuron as a child, you'll want to rediscover this special place," by most accounts have not been disappointed. In the interim, MacGregor Park has flourished as "a haven to over 105 migratory birds, nesting warblers and songbirds," its varied forest habitat luring the Red-Bellied Woodpecker, Blue-Grey Gnatcatcher, and Carolina Wren to "extend their range northward,"[9] offering an opportunity during each spring migration for birders to also extend their southern Ontario range. Given that the controlled access zone around the Bruce Power generation site has become a chosen sanctuary for fauna, the area below the old shoreline between Tiverton and Port Elgin has returned substantially to its pre-nuclear patterns of non-human occupance. The fishery is another matter, a chapter in a larger sad Great Lakes story in which the non-native species entering as stowaways in the ballast waters of St. Lawrence Seaway vessels has played a part.[10]

Eugene Bourgois continues to act on his concern that neighbours of megaprojects have the information they need from large corporations that locate in their midst. In June 2008, in response to an inquiry from the Peace River Environmental Society, he travelled to Alberta at his own expense, touring in the Whitecourt, Grimshaw, and Falher areas northwest of Edmonton where Bruce Power proposes to build Alberta's first nuclear plant. Local people were impressed by his "grassroots" accounts of "what an ordinary person can be looking at when trying to deal with a big company."[11] Bourgois has served as a director of the Inverhuron and District Ratepayers Association, a one-hundred-member community group that has had a rocky history with the nuclear site. But by 2008, most local ratepayers were committed to maintaining improved communications with Bruce Power.[12]

The Walkerton case is different, more recent (not yet a decade in the past), spawned by a different kind of state colliding with inhabitants with a different ethos of citizenship. Theirs was in several senses an internal rather than an external challenge to embodied balance, the victims disordered by practices locally embraced rather than installed from afar. At this stage, locally endorsed state instruments are conspicuous parts of learning "to go on going on" in Walkerton. The report of the O'Connor Commission has been formative in the remaking of water governance internationally. The Walkerton Clean Water Centre provides courses for water operators, advice to the provincial Ministry of the Environment, and a demonstration site for leading-edge water treatment and distribution technologies. The long-term health study, though not the study the Concerned Walkerton Citizens would have wished, has documented the health histories of four thousand area residents and provided access to specialists through local clinics. Most who were ill have recovered. Those whose health remains compromised were children in 2000. The Walkerton tragedy occurred through a collision of natural events and human circumstances, at a time when citizens could and did make demands of the state, in a community where responsibility for the system failure was widely shared. Mistaken inferences from embodied knowledge had been the undoing of the town. Knowing anew in Walkerton has been and likely will continue to be multiply testing of local confidence in local ways.

Notes

All interviews, unless otherwise noted, were conducted by Joy Parr.

FOREWORD

1 F. R. Leavis and Denys Thompson, *Culture and Environment: The Training of Critical Awareness* (London: Chatto and Windus, 1933), 1. This is but one of a long series of works that have echoed the same fundamental theme. For a useful summary and overview, see Jonathan Bate, "Culture and Environment: From Austen to Hardy," *New Literary History* 30, 3 (1999): 541-60.

2 Raymond Williams, *Culture and Society, 1780-1950* (New York: Doubleday, 1960), 278; Gary Day, *Re-reading Leavis: "Culture" and Literary Criticism* (New York: Palgrave Macmillan, 1996), 47-75. See also John Storey, *Cultural Theory and Popular Culture: An Introduction* (Harlow, England; New York: Prentice Hall, 2001), 22-28.

3 Philip Marchand, *Marshall McLuhan: The Medium and the Messenger: A Biography* (Toronto: Vintage Canada, [1998] 1989), 40. Indeed, McLuhan's first major work, *The Mechanical Bride: Folklore of Industrial Man* (New York: Vanguard Press, 1951), grappled with topics central to *Culture and Environment* (as it echoed some of its concerns in its subtitle) by exploring the effects of advertising on society and culture.

4 Marshall McLuhan and Quentin Fiore, *The Medium is the Massage: An Inventory of Effects* (New York: Bantam Books, 1967), 63.

5 "Marshall McLuhan and the Senses," *MMcL Magazine*, Ginko Press, http://www.gingkopress.com.

6 Joy Parr, *Labouring Children: British Immigrant Apprentices to Canada, 1869-1924* (London: Croom Helm; Montreal: McGill-Queen's University Press, 1980); Joy Parr, *The Gender of Breadwinners: Women, Men, and Change in Two Industrial Towns, 1880-1950* (Toronto: University of Toronto Press, 1990); Joy Parr, *Domestic Goods: The Material, the Moral, and the Economic in the Postwar Years* (Toronto: University of Toronto Press, 1999).

7 Mark M. Smith, "Making Sense of Social History," *Journal of Social History* 37, 1 (2003): 165-86, quote on 167.

8 Joseph Addison, "Pleasures Of Imagination," *Spectator*, no. 411, 21 June 1712.

9 Smith, "Making Sense of Social History," 166.

10 See Denis Cosgrove, *Social Formation and Symbolic Landscape* (Beckenham, Kent: Croom Helm, 1984); Denis Cosgrove, "Prospect, Perspective and the Evolution of the Landscape Idea," *Transactions of the Institute of British Geographers*, New Series, 10, 1 (1984): 45-62; and Denis Cosgrove, *Geographical Imagination and the Authority of Images* (Heidelberg: Department of Geography, 2006). For a brief but broad-ranging discussion, see John Wylie, "Landscape" in Derek Gregory, Ron Johnston, Geraldine Pratt, Michael J. Watts, and Sarah Whatmore, eds., *The Dictionary of Human Geography*, 5th edition (Chichester: Wiley-Blackwell, 2009), 409-11. The substance of this paragraph is expanded upon in many places, including, for example, in Chapter 7, "Visual Geographies," of Paul Rodaway's *Sensuous Geographies: Body, Senses and Place* (London; New York: Routledge, 1994), 115-44.

11 D.C.D. Pocock, "Sight and Knowledge," *Transactions of the Institute of British Geographers*, New Series, 6, 4 (1981): 385-93, quote on 385.

12 Richard Cavell, *McLuhan in Space: A Cultural Geography* (Toronto: University of Toronto Press, 2002), 70. See also Donald Theall, "McLuhan's Canadian Sense of Space, Time and Tactility," *Journal of Canadian Studies* 37, 3 (Fall 2002): 251-54.

13 Nancy Shaw, in a review of Richard Cavell, "*McLuhan in Space*, in *ESC: English Studies in Canada* 32, 2-3 (2008): 255-57, makes the interesting claim that "the primary intent of *McLuhan in Space* is to introduce concepts of acoustic space to critical geographers."

14 B.A. Kennedy, "A Naughty World," *Transactions of the Institute of British Geographers* 4 (1979): 550-58.

15 Michael Bull, Paul Gilroy, David Howes, and Douglas Kahn, "Introducing Sensory Studies," *Senses and Society* 1 (March 2006): 5-7.

16 George H. Roeder Jr., "Coming to Our Senses," *Journal of American History* 81 (December 1994): 1112-22, esp. 1112-13; Mark M. Smith, "Still Coming to "Our" Senses: An Introduction," *Journal of American History* 95, 2 (September 2008): 378-79.

17 A few recent examples of American histories on this theme include Leigh Schmidt, *Hearing Things: Religion, Illusion, and the American Enlightenment* (Cambridge, MA: Harvard University Press, 2000); Roger Horowitz, *Putting Meat on the American Table: Taste, Technology, Transformation* (Baltimore: Johns Hopkins University Press, 2006); Mark M. Smith, *How Race Is Made: Slavery, Segregation, and the Senses* (Chapel Hill: University of North Carolina Press, 2006); Mark M. Smith, *Mastered by the Clock: Time, Slavery, and Freedom in the American South* (Chapel Hill: University of North Carolina Press, 1997); and Mark M. Smith, *Listening to Nineteenth-Century America* (Chapel Hill: University of North Carolina Press, 2001). For a general survey, see also Mark M. Smith, *Sensing the Past: Seeing, Hearing, Smelling, Tasting, and Touching in History* (Berkeley: University of California Press, 2008). For examples from environmental history and history of science, see Linda Nash, *Inescapable Ecologies: A History of Environment, Disease, and Knowledge* (Berkeley: University of California Press, 2007); Conevery Bolton Valenčius, *The Health of the Country: How American Settlers Understood Themselves and Their Land* (New York: Basic Books, 2002); Gregg Mitman, *Breathing Space: How Allergies Shape Our Lives and Landscapes* (New Haven: Yale University Press, 2007); Gregg Mitman, Michelle Murphy, and Christopher Sellers, eds., "Landscapes of Exposure: Knowledge and Illness in Modern

Environments," special issue, *Osiris* 19 (2004); and Gregg Mitman, "In Search of Health: Landscape and Disease in American Environmental History," *Environmental History* 10 (2005): 184-209.

18 Stephen Mosley, "Common Ground: Integrating Social and Environmental History," *Journal of Social History* 39, 3 (2006): 915-33; Alan Taylor, "Unnatural Inequalities: Social and Environmental Histories," *Environmental History* 1 (1996): 6-19; Jeffrey K. Stine and Joel A. Tarr, "At the Intersection of Histories: Technology and the Environment," *Technology and Culture* 39 (1998): 601-40; Ted Steinberg, "Down to Earth: Nature, Agency, and Power in History," *American Historical Review* 107 (2002): 798-820.

19 Note here David Howes, ed., *Empire of the Senses: The Sensual Culture Reader* (Oxford: Berg, 2004); David Howes, *Sensual Relations: Engaging the Senses in Culture and Social Theory* (Ann Arbor, MI: University of Michigan Press, 2003); and David Howes, ed., *The Varieties of Sensory Experience: A Sourcebook in the Anthropology of the Senses* (Toronto: University of Toronto Press, 1991).

20 Lucien Febvre, *Le problème de l'incroyance au XVIe siècle : La religion de Rabelais* (Paris: Éditions Albin Michel, 1947), trans. Beatrice Gottlieb, *The Problem of Unbelief in the Sixteenth Century: The Religion of Rabelais* (Cambridge, MA: Harvard University Press 1982); Robert Mandrou, *Introduction à la France Moderne : Essai de psychologie historique 1500-1640* (Paris: Éditions Albin Michel, 1961), trans. R.E. Hallmark, *Introduction to Modern France, 1500-1640: An Essay in Historical Psychology* (New York: Holmes and Meier, 1976).

21 Jacques le Goff, "Au Moyen Age: Temps de L'Eglise et temps du marchand," *Annales : Économies, Sociétés, Civilisations* 15 (1960): 417-33.

22 Alain Corbin, *Le temps, le désir et l'horreur : Essais sur le dix-neuvième siècle* (Paris: Flammarion, 1991), trans. Jean Birrell, *Time, Desire and Horror: Towards a History of the Senses* (Cambridge: Polity Press, 1995); Alain Corbin, *Le Miasme et la jonquille: l'odorat et l'imaginaire social, 18-19e siècles* (Paris: Flammarion, 1986), trans. Miriam Kochan, Roy Porter, and Christopher Prendergast, *The Foul and the Fragrant: Odor and the French Social Imagination* (Cambridge, MA: Harvard University Press, 1986); Alain Corbin, *Les cloches de la terre : paysage sonore et culture sensible dans les campagnes au XIXe siècle* (Paris: A. Michel, 1994), trans. Martin Thom, *Village Bells: Sound and Meaning in the 19th-Century French Countryside* (New York: Columbia University Press, 1998). See for a commentary Sima Godfrey, "Alain Corbin: Making Sense of French History," *French Historical Studies* 25 (2002): 381-98.

23 E.P. Thompson, "Time, Work-Discipline, and Industrial Capitalism," *Past and Present* 38 (December 1967): 56-97; Smith, "Making Sense of Social History," 167.

24 Ted Hughes, *Tales from Ovid: Twenty-four Passages from the "Metamorphoses"* (New York: Farrar Straus and Giroux, 1997).

25 I adapt this phrase deliberately from Anne Michaels' luminous novel *The Winter Vault* (Toronto: McClelland and Stewart, 2009), 109, in which one of her characters contemplates the new houses – "hollow blocks of concrete" with no connection to the ground – built for Nubians displaced by the High Aswan dam, recognizing that "life can be skinned of meaning, skinned of memory"; the story also turns, in substantial part, on the displacement and dislocation caused by the construction of the St. Lawrence Seaway, one of Parr's foci in this book.

26 Quoted in Graeme Wynn, *Canada and Arctic North America: An Environmental History* (Santa Barbara: ABC-Clio, 2007), 289.

27 Paul R. Josephson, *Industrialized Nature: Brute Force Technology and the Transformation of the Natural World* (Washington: Island Press, 2002).
28 Donald Worster, "Two Faces West: The Development Myth in Canada and the US," in Paul W. Hirt, ed. *Terra Pacifica: People and Place in the Northwest States and Western Canada* (Pullman: Washington State University Press, 1998), 86-87.
29 Charles Bruce, "Orchard in the Woods," in *The Mulgrave Road: Selected Poems of Charles Bruce* (Toronto: Macmillan, 1951).
30 Roderick Frazier Nash, "Blurb," on John Warfield Simpson, *Yearning for the Land: A Search for the Importance of Place* (New York: Pantheon Books, 2002).

THE MEGAPROJECTS NEW MEDIA SERIES

1 This database-related aspect of new media works is discussed in Joy Parr, Jon van der Veen, and Jessica Van Horssen, "The Practicing of History Shared across Differences: Needs, Technologies and Ways of Knowing in the Megaprojects New Media Project," *Journal of Canadian Studies* 43, 1 (2009): 48-58.
2 Diverse examples of interesting work within the database model of historical new media scholarship include: Roy Rosenzweig, "Can History be Open Source? Wikipedia and the Future of the Past," in *The Journal of American History* 93, 1 (June, 2006): 117-46; Daniel Cohen, "From Babel to Knowledge: Data Mining Large Digital Collections," *D-Lib Magazine* 12, 3 (March, 2006): 6-19; Michael Frisch, "Oral History and the Digital Revolution: Towards a Post-Documentary Sensibility," in Robert Perks and Alistair Thompson, eds., *The Oral History Reader*, 2nd ed. (New York: Routledge, 2006), 102-14. See also Concordia University's Digital History Lab at http://digitalhistory.concordia.ca.
3 Phenomenological approaches to new media are relevant in this endeavour, including: Mark Hansen, *Bodies in Code: Interfaces with Digital Media* (London: Routledge, 2006); Laura Marks, *Touch: Sensuous Theory and Multisensory Media* (Minneapolis: University of Minnesota Press, 2002).
4 See http://megaprojects.uwo.ca/Iroquois and http://megaprojects.uwo.ca/Gagetown. The exception here is the Walkerton project at http://megaprojects.uwo.ca/Walkerton, which, as discussed in its introduction, is of different progeny.

ACKNOWLEDGMENTS

1 See "Lives Lived," *Globe and Mail*, 18 July 2000.
2 This was not quite so linear a process as I have summarized there. See Joy Parr, Jon van der Veen, and Jessica Van Horssen, "The Practicing of History Shared across Differences: Needs, Technologies and Ways of Knowing in the Megaprojects New Media Project," *Journal of Canadian Studies* 43, 1 (2009): 1-24.

CHAPTER 1: INTRODUCTION

1 A field of inquiry dedicated specifically to these relationships has been pursued since the early 1990s by David Howes and Constance Classen at Concordia University. See the journal entitled *The Senses and Society*, edited by Howes, and the companion monograph series on these themes. Both are published by Berg. Mark Smith's *Sensory History: An Introduction* (Oxford: Berg, 2007) appeared first in this series and was later published as

Sensing the Past: Seeing, Hearing, Smelling, Tasting and Touching in History (Berkeley: University of California Press, 2008). A special issue on sensory history appeared in the *Journal of American History* in October 2008.

2 I borrow this useful concept from Barbara Duden, *Women beneath the Skin: A Doctor's Patients in Eighteenth-Century Germany* (Cambridge, MA: Harvard University Press, 1991).

3 Boris Cyrunik, *The Whispering of Ghosts* (New York: Other Press, 2005); Steve Pile, "Ghosts and the City of Hope," in Loretta Lees, ed., *The Emancipatory City? Possibilities and Paradoxes* (London: Sage, 2004), 210-28; Michel de Certeau and Luce Giard, "Ghosts in the City," in *The Practice of Everyday Life*, vol. 2, *Living and Cooking*, trans. Timothy J. Tomsik (Minneapolis: University of Minnesota Press, 1998), 133-44.

4 Mark Hansen, *Embodying Technesis: Technology beyond Writing* (Ann Arbor: University of Michigan Press, 2000).

5 N. Katherine Hayles, "Foreword: Clearing the Ground," vii, in Hansen, *Embodying Technesis*. There are case studies within environmental history. See Peter Coates, "The Strange Stillness of the Past: Toward an Environmental History of Sound and Noise," *Environmental History* 10 (2005): 656; Joy Parr, "Smells Like? Sources of Uncertainty in the History of the Great Lakes Environment," *Environmental History* 11 (2006): 282-312.

6 Elaine Scarry, *Resisting Representation* (Oxford: Oxford University Press, 1994).

7 See http://megaprojects.uwo.ca/Gagetown.

8 The term was coined by Jon van der Veen, who joined this project while a master's student in media studies at the University of Western Ontario, and who continued, during his doctoral studies in communications at Concordia University in Montreal, to take the lead, designing and implementing our attempts to more thoroughly communicate the embodied transformations upon which my documentary and oral history work focused. See his "Lost in Transcription: Oral and Textural Readings of Interviews from Bruce County," (ms).

9 See http://megaprojects.uwo.ca/nuclear.

10 See http://megaprojects.uwo.ca/Iroquois.

11 See http://megaprojects.uwo.ca/ArrowLakes.

12 See http://megaprojects.uwo.ca/nuclear.

13 See http://megaprojects.uwo.ca/Walkerton.

14 Pierre Bourdieu, *The Logic of Practice* (Cambridge: Polity, 1990), 56.

15 Karl Marx, "Private Property and Communism," in D.J. Struick, ed., *The Economic and Philosophic Manuscripts of 1844* (New York: International Publishers, 1972), 133-46. The quote is at 141.

16 Henri Lefebvre, *The Production of Space* (Oxford: Blackwell 1991), 211, 195, 200, 175-77.

17 René Descartes, *The Philosophical Writings of René Descartes*, vol. 2, *Meditations on First Philosophy*, trans. J. Cottingham, R. Stoothoff, and D. Murdoch (Cambridge: Cambridge University Press, 1984 [1641]), 2:1-62; Joan Scott, "The Evidence of Experience," *Critical Inquiry* 17 (1991): 773-97, and *Gender and the Politics of History* (New York: Columbia University Press, 1988).

18 N. Katherine Hayles, "Situated Nature and Natural Simulations: Rethinking the Relation between the Beholder and the World," in William Cronon, ed., *Uncommon Ground: Toward Reinventing Nature* (New York: Norton, 1995), 409-25, and her "Searching for Common Ground," in Michael E. Soulé and Gary Lease, eds., *Reinventing Nature? Responses to Postmodern Deconstruction* (Washington, DC: Island Press, 1995), 46-63.

19 Hayles, "Searching for Common Ground," 48.

20 See Edmund Russell, "Evolutionary History: Prospectus for a New Field," *Environmental History* 8 (April 2003): 205.

21 Edward T. Hall, *The Hidden Dimension* (Garden City, NY: Doubleday, 1966); and Hall, *The Silent Language* (Garden City, NY: Doubleday, 1959).

22 Ruth Schwartz Cowan, *More Work for Mother: The Ironies of Household Technology from the Open Hearth to the Microwave* (New York: Basic Books, 1983); Susan Strasser, *Never Done: A History of American Housework* (New York: Pantheon, 1982).

23 Michael Polanyi, "Tacit Knowing," in Polanyi, *The Tacit Dimension* (Garden City, NY: Doubleday, 1966), 3-25.

24 Marcel Mauss, "Les techniques du corps," *Journal de psychologie (sociologie et anthropologie)* 32 (1936): 363-86; Pierre Bourdieu, *Outline of a Theory of Practice* (Cambridge: Cambridge University Press, 1977), chap. 2, 78-87.

25 Hansen, *Embodying Technesis,* 27.

26 William Cronon, "A Place for Stories: Nature, History and Narrative," *Journal of American History* 78 (March 1992): 1347-76.

27 Here, I am borrowing the terms of Hansen, *Embodying Technesis,* 26-27.

28 For example, the embodiment studied by Simon Newman in *Embodied History: The Lives of the Poor in Early Philadelphia* (Philadelphia: University of Pennsylvania Press, 2003) is not corporeal but epistemological. It is about discourses rather than material bodies.

29 Barbara Duden, *Women beneath the Skin,* 14-15, 35-7, 79-80; Barbara Duden, *Disembodying Women: Perspectives on Pregnancy and the Unborn* (Cambridge, MA: Harvard University Press, 1993), 91-92; Isabel V. Hull, "The Body as Historical Experience: Review of Recent Works by Barbara Duden," *Central European History* 28 (1995): 75; Elaine Scarry, *The Body in Pain: The Making and Unmaking of the World* (Oxford: Oxford University Press, 1985); and Scarry, *Resisting Representation* (Oxford: Oxford University Press, 1994).

30 Stanley J. Reiser, "Technology and the Use of the Senses in Twentieth-Century Medicine," in W.F. Bynum and Roy Porter, eds., *Medicine and the Five Senses* (New York: Cambridge University Press, 1993), 262-73.

31 Marshall McLuhan, *Gutenberg Galaxy: The Making of Typographic Man* (Toronto: University of Toronto Press, 1962); Walter J. Ong, "The Shifting Sensorium," in *The Presence of the Word* (New Haven: Yale University Press, 1967), 1-9.

32 Elizabeth Grosz, "Bodies and Knowledges," in Grosz, *Space, Time and Perversion: Essays on the Politics of Bodies* (London and New York: Routledge, 1995), 31; Leslie Adelson, *Making Bodies, Making History: Feminism and German Identity* (Lincoln: University of Nebraska Press, 1993), 12, 16, 23.

33 Alain Corbin, *The Foul and the Fragrant: Odor and the French Social Imagination* (Cambridge, MA: Harvard, 1986 [1982]); Corbin, *Village Bells: Sound and Meaning in the Nineteenth-Century French Countryside* (New York: Columbia University Press, 1998 [1994]); Corbin, *The Lure of the Sea: The Discovery of the Seaside in the Western World, 1750-1840* (Oxford: Blackwell, 1994 [1988]); Corbin, *Histoire du corps* (Paris: Seuil, 2005).

34 Sima Godfrey, "Alain Corbin: Making Sense of French History," *French Historical Studies* 25 (2002): 395-96; Alain Corbin, *The Life of an Unknown: The Rediscovered World of a Clogmaker in Nineteenth-Century France* (New York: Columbia University Press, 2001).

35 Humberto Maturana and Francisco Varela, *Tree of Knowledge: The Biological Roots of Human Understanding* (Boston: Shambhala, 1992), 24.

36 Adelson, *Making Bodies*, 3, 10.

37 Humberto Maturana and Bernhard Poerksen, *From Being to Doing: The Origins of the Biology of Cognition* (Heidelberg: Carl-Auer, 2004), 16-17; and Maturana and Varela, *Tree of Knowledge*, 24.

38 Hayles, "Searching for Common Ground," 49-50.

39 Polanyi, "Tacit Knowing"; Hayles, "Searching for Common Ground," 49-50; Edward S. Casey, "How to Get from Space to Place in a Fairly Short Stretch of Time," in Steven Feld and Keith Basso, eds., *Senses of Place* (Santa Fe, NM: School of American Research Press, 1996), 18-19.

40 Scott, "The Evidence of Experience"; and Scott, *Gender and the Politics of History*.

41 Paul Connerton, *How Societies Remember* (New York: Cambridge University Press, 1989); Francisco Varela, Evan Thompson, and Eleanor Rosch, *The Embodied Mind: Cognitive Science and Human Experience* (Cambridge, MA: MIT Press, 1991); Maurice Merleau-Ponty, *Phenomenology of Perception* (New York: Humanities Press, 1962); Pierre Bourdieu, *Outline of a Theory of Practice* (New York: Cambridge University Press, 1977).

42 Laura Gowing, *Women, Touch and Power in Seventeenth-Century England* (New Haven: Yale University Press, 2003) is a compelling exploration of what was held private and what became public in sense-making about giving birth.

43 Kathleen Canning, "The Body as Method? Reflections on the Place of the Body in Gender History," in Canning, *Gender and History in Practice: Historical Perspectives on Bodies, Class and Citizenship* (Ithaca: Cornell University Press, 2005), 168.

44 Varela, Thompson, and Rosch, *The Embodied Mind*; and Evan Thompson, "The Mindful Body: Embodiment and Cognitive Science," in Michael O'Donovan-Anderson, ed., *The Incorporated Self: Interdisciplinary Perspectives on Embodiment* (Lanham, MD: Rowman and Littlefield, 1996).

45 Douglas Harper, *Working Knowledge: Skill and Community in a Small Shop* (Chicago: University of Chicago Press, 1987). He is working with the insights first articulated by Mauss as bodily techniques. See Mauss, "Les techniques du corps."

46 Adam Rome, "From the Editor," *Environmental History* 9 (2004): 6-7.

47 Peter Coates, "The Strange Stillness of the Past: Toward an Environmental History of Sound and Noise," *Environmental History* 10 (2005): 656; and Joy Parr, "Smells Like? Sources of Uncertainty in the History of an Environment," *Environmental History* 11 (2006): 282.

48 N. Katherine Hayles, "Foreword: Clearing the Ground," vii, in Hansen, *Embodied Technesis*.

49 This phrase is Isabel V. Hull's praise for Barbara Duden's practice in her "The Body as Historical Experience," 75.

50 Joe Corn, "'Textualizing Technics': Owner's Manuals and the Reading of Objects," in Ann Smart Martin and J. Ritchie Garrison, eds., *American Material Culture: The Shape of the Field* (Winterthur, DE: Winterthur Museum, 1997), 169-94.

51 Barbara L. Allen, "Narrating the Toxic Landscape in 'Cancer Alley,' Louisiana," in David Nye, ed., *Technologies of Landscape: From Reaping to Recycling* (Boston: University of Massachusetts Press, 1999), 188-89.

52 Allen, "Narrating the Toxic Landscape," 193, 195, 199. This is knowledge that the cottagers, campers, and pastoralists whom we meet in Chapter 6 lacked, and which they knew they lacked. It is knowledge that the radiation protection instruction for nuclear workers (analyzed in Chapter 3) sought to convey.

53 Miriam Wright, "Young Men and Technology: Government Attempts to Create a 'Modern' Fisheries Workforce in Newfoundland, 1949-1970," *Labour/Le travail* 42 (1998): 143-59.

54 Thomas J. Csordas, "Modes of Somatic Attention," *Cultural Anthropology* 8 (1993): 135-56.

55 Raymond Smilor, "Personal Boundaries in the Urban Environment: The Legal Attack on Noise, 1865-1930," in Char Miller and Hal Rothman, eds., *Out of the Woods: Essays in Environmental History* (Pittsburgh: University of Pittsburgh Press, 1997), 181-93.

56 Virginia Scharff, "Lighting Out for the Territory: Women, Mobility and Western Place," in Richard White and John Findlay, eds., *Place and Power in the North American West* (Seattle: University of Washington Press, 1999), 287-303; Scharff, "Of Parking Spaces and Women's Places: The Los Angeles Parking Ban of 1920," *National Women's Studies Association Journal* 1 (1988): 37-51; Scharff, *Taking the Wheel: Women and the Coming of the Motor Age* (New York: Free Press, 1991).

57 Hansen, *Embodying Technesis*, vi, 1-2.

58 Rebecca Solnit, *Wanderlust: A History of Walking* (New York: Viking, 2000); and Joseph Amato, *On Foot: A History of Walking* (New York: New York University Press, 2004). My own research on walking is reported in Chapter 4. See also Jon van der Veen's walking tours at http://megaprojects.uwo.ca/Iroquois.

59 Hayden White, *The Content of the Form: Narrative Discourse and Historical Representation* (Baltimore: Johns Hopkins University Press, 1987); and Marshall McLuhan, *Understanding Media: The Extensions of Man* (New York: McGraw-Hill, 1964).

60 David Nye's writing about origin stories makes similar points. But for our purposes here, I've decentred the narrative. See David Nye, "Technology, Nature and American Origin Stories," *Environmental History* 8 (2003): 8-24.

61 Thomas R. Dunlap, "Australian Nature, European Culture: Anglo Settlers in Australia," in Char Miller and Hal Rothman, eds., *Out of the Woods: Essays in Environmental History* (Pittsburgh: University of Pittsburgh, 1997), 273-89.

62 Carolyn Merchant, "Reinventing Eden: Western Culture as a Recovery Narrative," in William Cronon, *Uncommon Ground: Toward Reinventing Nature* (New York: Norton, 1995), 132, 133, 144, 158. See also Carolyn Merchant, "Gender and Environmental History," *Journal of American History* 76 (1990): 1117, 1119; Merchant, *Ecological Revolutions: Nature, Gender and Science in New England* (Chapel Hill: University of North Carolina Press, 1989); Merchant, *Death of Nature: Women, Ecology and the Environmental Revolution* (San Francisco: Harper and Row, 1980); Vera Norwood, *Made from the Earth: American Women and Nature* (Chapel Hill: University of North Carolina Press, 1993); Max Oelschlaeger, "Re-Placing History, Naturalizing Culture," in John P. Herron and Andrew G. Kirk, eds., *Human/Nature: Biology, Culture and Environmental History* (Albuquerque: University of New Mexico, 1999), 69.

63 William Cronon, "The Trouble with Wilderness: Or, Getting Back to the Wrong Nature," in Miller and Rothman, *Out of the Woods*, 46-49.

64 A stance in practice and interpretive promise well recounted in Evelyn Fox Keller, *A Feeling for the Organism: The Life and Work of Barbara McClintock* (San Francisco: W.H. Freeman, 1983).

65 Judith Butler, *Gender Trouble: Feminism and the Subversion of Identity* (New York: Routledge, 1990); and Butler, *Bodies That Matter: On the Discursive Limits of Sex* (New York: Routledge, 1993).

66 Joan Scott, "Gender: A Useful Category of Historical Analysis," *American Historical Review* 91 (1986): 1053-75; and Scott, *Gender and the Politics of History* (New York: Columbia University Press, 1988).

67 Rachel Carson, *Silent Spring* (Greenwich, CT: Fawcett, 1962), 261; Oelschlaeger, "Re-Placing History, Naturalizing Culture," 63-78; Maril Hazlett, "'Woman vs. man vs. bugs': Gender and Popular Ecology in Early Reactions to Silent Spring," *Environmental History* 9 (2004): 701-29.

68 Mauss, "Les techniques du corps," 363-86.

69 Vera Norwood, "Constructing Gender," in Herron and Kirk, *Human/Nature*, 50, drawing upon Hayles, "Searching for Common Ground."

70 Norwood, "Constructing Gender," 55, 57, 60.

71 Paul Mohai, "Men, Women and the Environment: An Examination of the Gender Gap in Environmental Concern and Activism," *Society and Natural Resources* 5 (1992): 1-19.

72 Phil Brown and Faith T. Ferguson, "'Making a Big Stink': Women's Work, Women's Relationships and Toxic Waste Activism," *Gender and Society* 9 (1995): 145-72.

73 Dorothy Smith, *The Everyday World as Problematic: A Feminist Sociology* (Boston: North-eastern University Press, 1987), 51; Sara Ruddick, *Maternal Thinking: Towards a Politics of Peace* (Boston: Beacon, 1989), 70.

74 Gregg Mitmann, *Breathing Space: How Allergies Shape Our Lives and Landscapes* (New Haven: Yale University Press, 2007), 254.

75 Linda Nash, *Inescapable Ecologies: A History of Environment, Disease and Knowledge* (Berkeley: University of California Press, 2006), 208, 209, 215; and Nash, "Finishing Nature: Harmonizing Bodies and Environments in Late-Nineteenth-Century California," *Environmental History* 8 (2003): 6, 15.

76 Conevery Bolton Valenčius, *The Health of the Country: How American Settlers Understood Themselves and Their Land* (New York: Basic Books, 2002), 3, 12-13, 22, 53, 85, 163, 199, 225-27.

CHAPTER 2: PLACE AND CITIZENSHIP

1 Marion Reicker, *Those Days Are Gone Away: Queens County, N.B., 1643-1901* (Gagetown: Queens County Historical Society, 1981), 93.

2 Walter Benjamin, *Illuminations* (New York: Harcourt Brace, 1968), 249.

3 Two recent books situate this displacement: James Laxer, within the development of French-English relations in New Brunswick and Canada, *The Acadians in Search of a Homeland* (Toronto: Doubleday, 2006); and John Mack Faragher, within histories of ethnic cleansing, *A Great and Noble Scheme: The Tragic Story of the Expulsion of the French Acadians from Their American Homeland* (New York: Norton, 2005).

4 "Move to Obtain Brigade Training Camp in District," *Telegraph-Journal*, Saint John, 15 May 1951; "Canada Has Not Space to Train Her Own Soldiers," *Telegraph-Journal*, 11 July 1951; "Canada Requires Large Training Area," *Fredericton Gleaner*, 11 July 1951; "Utopia Site Meeting Slated for Tonight," *Fredericton Gleaner*, 28 September 1951; "To Build Largest Army Base in NB," *Telegraph-Journal*, 16 July 1952; "Canada's Biggest Army Base," *Telegraph-Journal*, 17 July 1952; "Text of Claxton Statement," *Fredericton Gleaner*, 1 August 1952; "Training Camp for 10,000 Men Starts at Camp Gagetown," *Telegraph-Journal*, 5 July 1955; "Army Training Base Main Heating Plant to Burn Minto Coal," *Telegraph-Journal*, 22 April 1955, 1, 10.

5 "Decision on Training Camp Location Expected in a Few Weeks," *Telegraph-Journal,* 28 September 1951; "Huge NB Army Campsite Is Confirmed," *Telegraph-Journal,* 2 August 1952; "Picture of Arm's Camp at Gagetown Is Given in Detail," *Telegraph-Journal,* 12 February 1954; "Army Says Gagetown Negotiations Were Relations Triumph," 16 April 1955; "Training Camp for 10,000 Men," *Telegraph-Journal,* 5 July 1955; Queens County Historical Society and Museum (hereafter QCHSM), CFB Gagetown Expropriation Documentation, 488-2825, Lydia Clarke Scott Scrapbooks, vol. 2; Gordon Swartzen, "The Impact of a Military Installation on a Local Economy" (MA thesis, University of Ottawa, 1963), chap. 1.

6 Carman Miller, "The 1940s: War and Rehabilitation," in E.R. Forbes and D.A. Muise, eds., *The Atlantic Provinces in Confederation* (Toronto: University of Toronto Press, 1993), 336, 384-85.

7 "Administrative Area for Vast New Army Camp," *Telegraph-Journal,* 23 July 1952; "Camp to Bring Heavy Income," *Telegraph-Journal,* 16 July 1952; "Hundreds Hear Army Plans," *Telegraph-Journal,* 19 August 1952; "May Not Be Forced to Move for 3 Years," *Telegraph-Journal,* 22 August 1952; "What Price Do They Pay for Contentment," *Fredericton Gleaner,* 13 September 1952; Public Archives of New Brunswick (hereafter PANB), MC 1558, Dorothy Dearborn Papers, records from the 1983 30th Camp Gagetown Reunion; PANB, MC 2926, interview of Rev. Robert Johnson by Jeff McCarthy, 27 August 2000, recounting that his father, "quite a royalist," would not have rebelled at a request made in the name of the queen.

8 "Gagetown-Welsford Military Training Area Opposed at Meeting," *Fredericton Gleaner,* 4 July 1952; "Heavy Opposition to Expected Base Site," *Fredericton Gleaner,* 19 July 1952; "Assures NB Claims for Training Centre Constantly Pressed," *Telegraph-Journal,* 3 October 1951; "Royal Member Opposed Queen's Army Camp Site," 21 August 1952.

9 PANB, MC 1558, Dearborn Papers, file 3, Petition dated July 1952; and QCHSM, 488-2825, Scott Scrapbooks, vol. 2; interview of David McKinney, Fredericton, NB, 26 October 2003.

10 "See Rapid Westfield Expansion as Heart of Army Training Camp," *Telegraph-Journal,* 23 July 1952; "Two Sides to Every Picture," *Telegraph-Journal,* 29 July 1952; PANB, MC 1558, file 3, letter circulated during the 1952 election campaign. In their subsequent meetings in Ottawa and Fredericton, the leaders of the District Farmers Association emphasized that their agricultural lands were as good as any in the province and that they were being asked to abandon other vital economic resources, among them their co-operative creamery at Hampstead and the managed stands of timber upon which they had relied for fuel and cash income.

11 "Are Given Promise of Consideration," *Telegraph-Journal,* 10 November 1952; PANB, MC 1558, file 3, report of a meeting in Fredericton, 3 December 1952; PANB, ref. 54334, side B, interview of Lydia Clarke Scott by Dorothy Dearborn, 21 June 1993; QCHSM, 488-2825, Scott Scrapbooks, vol. 4, Scripts of the CFBC broadcasts prepared by District Farmers Association No. 6, 13-18 October 1955.

12 Graeme Wynn, *Timber Colony: A Historical Geography of Early Nineteenth Century New Brunswick* (Toronto: University of Toronto Press, 1981), 156.

13 Normand Séguin, "L'économie agro-forestiére," *Revue de l'histoire de l'Amérique français* 29, 4 (1976) 559-65; René Hardy et Normand Séguin, *Forêt et société en Mauricie* (Québec: Laval, 2004).

14 Joseph Richards Petrie, *The Regional Economy of New Brunswick* (Fredericton: Canadian Committee on Reconstruction Report, 1943).

15 A.R.M. Lower, *Settlement and the Forest Frontier in Eastern Canada* (Toronto and New Haven: Ryerson and Yale University Press, 1938), 31-37.

16 Wynn, *Timber Colony,* 19, 83; Wynn, "Deplorably Dark and Demoralised Lumberers?" *Journal of Forest History* 24 (1980): 168-87; L. Anders Sandberg, *Trouble in the Woods: Forest Policy and Social Conflict in Nova Scotia and New Brunswick* (Fredericton: Acadiensis Press, 1992), 17-18; William Parenteau, "Forest and Society in New Brunswick: The Political Economy of Forest Industries" (PhD diss., University of New Brunswick, 1994), 6-10; R. Peter Gilles and Thomas R. Roach, *Lost Initiatives: Canada's Forest Industries – Forest Policy and Conservation* (New York: Greenwood, 1986), chap. 7; Thomas Roach, "Farm Woodlots and Pulpwood Exports from Eastern Canada, 1900-1930," in Harold K. Steen, ed., *History of Sustained-Yield Forestry: A Symposium* (Portland, OR: Forest History Society for the International Union of Forestry Research Organizations, 1984), 206.

17 Franklin Johnson, "Community of New Jerusalem," in Verna Mott, ed., *New Jerusalem: We Remember* (Gagetown: Base Gagetown History Association 2003), 28; Wynn, *Timber Colony,* 14-16; Matthew Betts and David Coon, *Working with the Woods: Restoring Forests and Community in New Brunswick* (Fredericton: Conservation Council of New Brunswick, 1996), 2-4.

18 D. Dyke and F. Lawrence, "Relocation Adjustments of Farm Families," *Economic Annalist,* February 1960, 5. The study was conducted in 1956 by the Rural Sociology Unit of the Economics Division of the federal Department of Agriculture and focused on 104 of the families who had been displaced. The authors reported that, formerly, commercial farmers had received 59 percent of their net cash income from the sale of agricultural products, 25 percent from their woodlots, and 8 percent from leased stumpage. Eighty-eight of the families had no indebtedness in 1952.

19 A provincial comparison makes this pattern clear. Whereas 57.6 percent of the occupied land in the province was classified as woods in 1911 and 58.9 percent in 1951, in the parishes of Gagetown, Hampstead, and Petersville of Queens County, woods covered 63 percent of the occupied land in 1911 and 67.2 percent in 1951. See *Census of Canada, Agriculture,* 1911, 1951.

20 Parenteau, "Forest and Society," 64; PANB, MC 1558, interview of Douglas Carr by Dorothy Dearborn, 11 November 1992.

21 The exception was the cutting and peeling of pulpwood, which happened in summer, between cropping and haying. See Parenteau, "Forest and Society," 73; Roach, "Farm Woodlots," 205.

22 There are several carefully told accounts of this process in the Base Gagetown Community Histories. I've relied here on David McKay, "Cutting Firewood," in Mott, *New Jerusalem: We Remember,* 78.

23 Dawn Bremner, ed., *Places of Our Hearts: Memories of Base Gagetown Communities to 1953* (Gagetown: Base Gagetown Community History Association and Queens County Historical Society and Museum, 2003), back cover; Connie Denby and Carol Lawson, eds., *Along Hibernia Roads: Faded Memories* (Gagetown: Base Gagetown Community History Association, 2003), 56, 109. Though the census does not provide information on car ownership at the subdistrict level relevant here, cars are commonly visible by garages and drivesheds that appear in the photographs taken by the teams evaluating properties at the time of expropriation.

24 "The Lumber Camps," in Denby and Lawson, *Along Hibernia Roads,* 121; Bremner, *Places of Our Hearts,* 61; Mott, *New Jerusalem: We Remember,* 26.

25 Austin Armstrong and Jennifer Walker, *Memories of Life in Two Counties* (Hampton, NB: privately published, 1998), 13; Betts, *Working in the Woods*, 12; *Census of Canada, Agriculture,* 1931, 1941, 1951.

26 Denby and Lawson, *Along Hibernia Roads*, 119, 121; PANB, MC 2926, interview of Lena McCann by Jeff McCarthy, 15 August 2000.

27 PANB, ref. 54334, interview of Lydia Clarke Scott by Dorothy Dearborn; Willard Clarke, "Byron Clarke's Mill," *Along Hibernia Roads*, 119-20.

28 Marion Gilchrist Reicker, *A Time There Was* (Gagetown: Queens County Historical Society, 1984), A-16.

29 Interview of Lydia Clarke Scott by Dorothy Dearborn. Similarly, on the woods as legacy rather than "windfall" and the elements of sustainable practice, see interview of Constance Denby, Base Gagetown, 27 October 2003; Bremner, *Places of Our Hearts,* 67; Denby and Lawson, *Along Hibernia Roads*, 110; PANB, MC 2969, interview of Margaret Ward by Jeff McCarthy, 16 August 2000; and interview of Lloyd McCann by Jeff McCarthy, 30 August 2000; QCHSM, Scott Scrapbooks, vol. 4, radio broadcast scripts, Raymond Scott, 14 October 1955. The Scotts' and Clarkes' land holdings are listed in Reicker, *A Time There Was*, A-73.

30 *Star Weekly,* 22 November 1952, 6; *Telegraph-Journal,* 23 July 1953.

31 "Biggest Individual Event, Says McNair," *Telegraph-Journal,* 2 August 1952 1, 5.

32 William Parenteau, "'In Good Faith': The Development of Pulpwood Marketing for Independent Producers in New Brunswick, 1960-1975," in Sandberg, *Trouble in the Woods,* 112.

33 The average income of commercial farms in the district was $2,645 in 1952, and a third of this income was derived from woodlots and stumpage. See Dyke and Lawrence, "Relocation Adjustments," 5.

34 QCHSM, Scott Scrapbooks, vol. 4, radio broadcast scripts, Raymond Scott, 14 October 1955; interview of Willard Clarke, Sussex, NB, 13 August 2003.

35 Some of the websites are listed at note 73.

36 Armstrong and Walker, *Memories of Life,* 19; Johnson, "Community of New Jerusalem," 24-26; "NB Dairymen's President Gives Reasons for Recent Increase in Butter Price," *Telegraph-Journal,* 9 November 1949, 2; "NB Dairy, Butter Associations Hold Annual Meetings At Sussex," *Telegraph-Journal,* 21 November 1951, 3; Norman Armstrong, "Armstrong Family History," in Richard Corbett, ed., *Summer Hill and Dunn's Corner: Gone But Not Forgotten* (Saint John, NB: Base Gagetown Community History Association, 2003), 29.

37 Wynn, *Timber Colony,* 156; interview of Lydia Clarke Scott by Dorothy Dearborn.

38 *Census of Canada, Agriculture*, 1901-51; Johnson, "Community of New Jerusalem," 24-26; Brian Donahue, *The Great Meadow: Farmers and the Land in Colonial Concord* (New Haven: Yale University Press, 2004), principally chaps. 3, 7, and 8, pp. 18, 55-56, 193, 206.

39 E.P. Reid and J. Fitzpatrick, *Atlantic Provinces Agriculture* (Ottawa: Canada Department of Agriculture, Economics Division, August 1957), 16.

40 There is a GIS series of the air photos, topographic sheets, and Google Earth images that illustrates these changes in the land use over time on http://megaprojects.uwo.ca/Gagetown.

41 Reid and Fitzpatrick note that the yields of field roots were thus higher than in Ontario and provided the main source of succulent feed for cattle. See *Atlantic Provinces Agriculture* 16; Dyke and Lawrence, "Relocation Adjustments," 5.

42 Hear Willard Clarke's song about pigs and cream on http://megaprojects.uwo.ca/Gagetown.

43 New Brunswick, Royal Commission on the Milk Industry in New Brunswick, *Report* 1970-71, 65-71.

44 Interview of Willard Clarke; interview of Lydia Clarke Scott by Dorothy Dearborn, 7; PANB MC 2926, interview of Murray Webb by Jeff McCarthy, 2 August 2002, Base Gagetown Community History Association.

45 RC on Milk, *Report* 67, 72; interview of Lydia Clarke Scott by Dorothy Dearborn, 2.

46 Interview of Constance Denby; "Dunn's Corner Herd Wins Jersey Club Silver Cup at West Royal Field Day," *Telegraph-Journal,* 28 June 1951; "Jersey Breeders Cancel Show," *Fredericton Gleaner,* 21 August 1952, 14.

47 "National Jersey Club Official at Gagetown," *Telegraph-Journal,* 29 March 1951, 2; "Queen's Fair Draws 3,500 during Second Day's Run," *Telegraph-Journal,* 13 September 1951; "Are Given Promise of Consideration," *Telegraph-Journal,* 10 November 1952, 5; "Modest 'Snowflake' Top Heifer Calf," *Fredericton Gleaner,* 28 May 1951, 5.

48 Raymond Scott took over from Eugene Morris as secretary treasurer and, with help from the district agriculturalist and Ken Green (one of his companions on the trip to meet Claxton in Ottawa), distributed the $77,607 assets of the co-op among its members. Interview of David McKinney; interview of Constance Denby; "NB Dairy, Butter Associations Hold Annual Meeting at Sussex," *Telegraph-Journal,* 21 November 1951; "Says Refectories of Greater Value to Queens County," *Telegraph-Journal,* 4 March 1953, 5; "Hampstead Co-operative Now Distributing Assets as Business Wound Up," *Telegraph-Journal,* 20 July 1954; Norman Armstrong, "Armstrong Family History," 29; and Isobel Johnston, "James Herbert McCullum and Mary Isabelle McCullum," in Corbett, *Summer Hill and Dunn's Corner,* 75; interview of Lydia Clarke Scott by Dorothy Dearborn, 3; RC on Milk *Report,* 33-36.

49 David Frank, "The 1920s," *Atlantic Provinces in Confederation,* 237; Donald Worster, "Thinking Like a River," in *The Wealth of Nations* (New York: Oxford University Press, 1993), 124-25; and, in the same volume, "The Transformation of the Earth," 48.

50 The number of farmers in 1931 was 484; in 1951 was 307: Census of Canada, *Agriculture* (1931, vol. 8, table 37, 182), (1951, vol. 6, table 29, 3); QCHSM, Scott Scrapbooks, vol. 4, radio broadcast script, 18 October 1955.

51 Douglas Harper, *Changing Works: Visions of a Lost Agriculture* (Chicago: University of Chicago Press, 2001) describes similar important reciprocities among the dairy farmers of northern New York. Eugene Morris is compelling on the "loss of neighbours who have stood with us in tough times and rejoiced with us in good times" and his justified apprehension that after the expropriation those displaced would have "no centre to return to when memory would call us to renew old friendships." See QCHSM, Scott Scrapbooks, vol. 4, broadcast script, 18 October 1955. Several photos on http://megaprojects.uwo.ca/ Gagetown show this propinquity.

52 Interview of Willard Clarke; interview of Lydia Clarke Scott by Dorothy Dearborn, 2; Bremner, *Places of Our Hearts,* 18, 21, 69, village ribbon in background picture 22; "Moving Up Front," *Telegraph-Journal,* 13 July 1954, 3, "deserted hamlet" of Summer Hill in background.

53 Dyke and Lawrence, "Relocation Adjustments," 5.

54 Bremner, *Places of Our Hearts,* 96, photo of New Brunswick Power Crew; interview of Lydia Clarke Scott by Dorothy Dearborn, 10; PANB, MC 2775, interview of Rev. Doug Woods by Erma Brian, 18 June 1998, 5.

55 "Says Persons Displaced by Camp 'Best Canadians There Are,'" *Telegraph-Journal*, 3 July
    1953, 3; "Farmers Seek Information on Army Camp," *Telegraph-Journal*, 3 November 1953.
    The estimate used by the Base Gagetown History Committee is 750 families and three
    thousand persons.

56 D. Dyke, "Relocation of 95 New Brunswick Farm Families, 1956-60," *Economic Annalist*,
    October 1962, 104.

57 Interview of Lydia Clarke Scott by Dorothy Dearborn, 8; Gerry Nickerson, "Great Scott,"
    in Corbett, *Summer Hill and Dunn's Corner*, 42-45; "Gagetown Festival Adjudicator
    Named," *Telegraph-Journal*, 16 February 1954, 2; interview of Maude Underhill, Gagetown,
    29 October 2003.

58 Interview of David McKinney; interview of Maude Underhill; interview of Willard Clarke.

59 *Star Weekly*, 22 November 1952, http://megaprojects.uwo.ca/Gagetown.

60 Interview of David McKinney; Johnson, "Community of New Jerusalem," 25; Ian Mac-
    donald, *Macdonalds at Thornhill Farm* (Perth, ON: privately published for Base Gagetown
    Reunion, 2003); Welsford Farm Forum, 13 February 1943; interview of Rev. Robert John-
    son by Jeff McCarthy, 8; interview of Franklin Johnson by Jeff McCarthy, 15 August 2000;
    "Aid Societies Hold Sessions," *Telegraph-Journal*, 3 October 1950, 3; "Queens-Sunbury
    Children's Aid Groups Elect Officers," *Telegraph-Journal*, 5 December 1952, 3; "Children's
    Aid Report Presented," *Telegraph-Journal*, 10 February 1953, 2; "Gagetown Sets Blood
    Clinic Date," *Fredericton Gleaner*, 25 October 1952, 3; "Gagetown Plans Fall Blood Clinic,"
    Telegraph-Journal, 25 October 1952, 3; Bremner, *Places of Our Hearts*, 73.

61 "Grand Orange Lodge Elected Officers," *Fredericton Gleaner*, 26 April 1951, 8; interview
    of David McKinney; Thomas Raymond Murphy, "The Structural Transformation of New
    Brunswick Agriculture from 1951 to 1981" (MA thesis, University of New Brunswick, 1983),
    271; "Prizes awarded in Bacon Contest," *Telegraph-Journal*, 15 May 1950, 3; "Queens
    Council Holds Meeting," *Telegraph-Journal*, 8 July 1952, 3; interview of Willard Clarke.

62 See W. Scott, *Riots in New Brunswick* (Toronto: University of Toronto Press, 1993), 27,
    78-81, 90, 93, 96-99, 105, 186-87.

63 Ibid., 88-89, 10.

64 PANB, MC 2775, interview of Rev. Doug Woods by Erma Brian, 18 June 1998; PANB,
    MC 2926, interview of Marian Mersereau by Jeff McCarthy, 22 August 2000; interview
    of Lloyd McCann; interview of Constance Denby.

65 The exception is Maude Underhill, who was uncomplimentary in her appraisal. Interview
    of Maude Underhill.

66 See http://megaprojects.uwo.ca/Gagetown.

67 Picture of Raymond Scott and Friends in Bremner, *Places of Our Hearts*, 71 and in Corbett,
    Summer Hill and Dunn's Corner, 40; interview of Constance Denby; see http://megaprojects.
    uwo.ca/Gagetown for work by Greg Marquis and Donn Downes of the University of New
    Brunswick, Saint John, with the fiddlers and dancers of the region.

68 "Proposed Vatican Appointment Irks Grand Orange Lodge," *Fredericton Gleaner*, 2 April
    1952, 3.

69 "Long Established Lodge Holds Colorful Parade Which May Be Its Last," *Telegraph-
    Journal*, 26 August 1952, 16; "Outdoor Service Held as Farewell to Orange Lodges,"
    Telegraph-Journal, 28 July 1953, 3.

70 Dyke and Lawrence, "Relocation Adjustments," 101-2, 109.

71 Interview of Willard Clarke.

72 "Call Bush Tenders for Camp Gagetown," *Telegraph-Journal*, 27 November 1954, 2; "Bulldozers and Cranes Take Over as Camp Gagetown Work Starts," *Telegraph-Journal*, 3 September 1953, 3; "Snow Holding Up Clearing of Land for Army Camp," *Telegraph-Journal*,
19 January 1955, 3; "Forest Fires Burning in Camp Gagetown Area," *Telegraph-Journal*, 23
May 1955, 2; "Keeping NB Forest Fires Under Control," *Telegraph-Journal*, 24 May 1955, 2.

73 National Defence and the Canadian Forces, "The Use of Herbicides at CFB Gagetown
from 1952 to the Present Day," 10 September 2006, http://www.forces.gc.ca/site/reports-
rapports/defoliant/index-eng.asp; Veterans Affairs Canada, "What Is Agent Orange?"
http://www.vac-acc.gc.ca; Sandra Williamson, Veterans Affairs Canada, "Agent Orange
Technical Briefing: CFB Gagetown, 23 June 2005," http://www.vac-acc.gc.ca; "Agent
Orange Health Risk to British Troops," 14 November 2006, http://www.gulfveteran-
sassociation.co.uk/DocumentArchive.htm; "Agent Orange Alert at CFB Gagetown," http://
www.agentorangealert.com

74 "Plans for Tank Training at Camp Start with Visit," *Telegraph-Journal*, 28 January 1954, 2;
"Advance Units Prepare for Troops," *Telegraph-Journal*, 1 June 1955 1, 10; "Tanks Fight
Gagetown Mud," *Telegraph-Journal*, 8 June 1954, 1, 5; interview of Rev. Robert Johnson
by Jeff McCarthy, 9; interview of Lena McCann by Jeff McCarthy, 13.

75 "Southern Section of Camp Gagetown Starting to Fill with Troops," *Telegraph-Journal*, 26
May 1954, 3; "Strong Counter-Thrust by 'Blueland' Forces," *Telegraph-Journal*, 11 August
1955, 3, 5; interview of Lena McCann by Jeff McCarthy, 13; interview of Lydia Clarke Scott
by Dorothy Dearborn, 7; "Canadian Troops Turn Deaf Ear to 'Surrender,'" *Telegraph-
Journal*, 3 August 1955, 3, 5; "Home-Made Mushroom at Camp Gagetown," *Telegraph-
Journal*, 19 July 1955, 3; "Whole Gagetown Area Changes as Army Moves in," *Daily
Standard Freeholder* (Cornwall), 16 August 1956, 11; PANB, MC 2926, interview of Mildred
Russell by Jeff McCarthy, 13 August 2000.

76 "Camp Gagetown Afford Good Fishing, Hunting," *Telegraph-Journal*, 24 June 1955, 20;
"Army Training Area Banned for Hunters," *Telegraph-Journal*, 4 October 1955, 2.

77 Swartzen, "The Impact of a Military Installation," 20, 30, citing Camp Gagetown Employment records.

78 "Fantasian Invitation Ignored," *Telegraph-Journal*, 3 August 1955, 3, 5; "'Philistia' Sends
Forces Marching into 'Blueland,'" *Telegraph-Journal*, 8 August 1955, 1, 2, 10; "Air Attack
Observed by Brigade Commander, Mascot," *Telegraph-Journal*, 11 August 1955, 3. Compare
image with http://www.virtualmuseum.ca, Base Gagetown Community History Associa-
tion, Summer Hill United Church of Canada, 1952; "Have Professional Army," *Telegraph-
Journal*, 17 June 1955, 10.

79 On the destruction of the structures, see "Past, Present and Future," *New Brunswick
Reader*, 26 July 2003, 16.

80 I began working around Gagetown in 2001, seeking help, and attended the 2003 reunion.
There is a photo of the baselands on the reunion weekend showing one of the signs identifying the site of a former building at http://megaprojects.uwo.ca/Gagetown.

81 Reicker, *Those Days Are Gone Away*, 196; interview of Maude Underhill; interview of David
McKinney; Will of Lydia Clarke Scott, probated 22 November 1996, Fredericton, NB;
Carman Miller, "The 1940s: War and Rehabilitation," in *Atlantic Provinces in Confedera-
tion*, ed. E.R. Forbes and D.A. Muise (Toronto: University of Toronto Press, 2001), 339.

82 "Ousted By Big Camp, Offered Shabby Farms," *Telegraph-Journal*, 17 June 1953, 3; "Displaced by Camp, Are Moving to Kings," *Telegraph-Journal*, 22 May 1953, 3; Dyke and

Lawrence, "Relocation Adjustments," 89; PANB, MC 1558, interview of Joe Day by Dorothy Dearborn, 1 November 1993.

83 Della Stanley, "The 1960s: The Illusions and Realities of Progress," 448; and John Reid, "The 1970s: Sharpening the Sceptical Edge," in *Atlantic Provinces in Confederation,* 483; Matthew Baglole, "Some of the People, Some of the Time: The Confederation of Regions Party in New Brunswick" (MA thesis, University of New Brunswick, 2002), 32, 117-18.

84 Interview of Lydia Clarke Scott by Dorothy Dearborn, 5, time 21:27.

85 QCHSM, Scott Scrapbooks, CORE no. 1, 1988-9, annotation on a New Brunswick Association of English-Speaking Canadians petition. Interview of David McKinney; interview of Maude Underhill; interview of Constance Denby. Baglole, "Some of the People," 5, 41, 50-51, 109, 134-35, 179. On immigration, Baglole cites Yasmeen Abu-Labab and Daiva Stasiulus, "Ethnic Pluralism Under Siege," *Canadian Public Policy* 18, 4 (1992): 365-86.

86 Interview of Maude Underhill; interview of David McKinney; interview of Willard Clarke. The designs, though not the naked ladies, often mirror drawings made by Lydia's students and preserved in one of her QCHSM, 488-2825 scrapbooks. Larry E. Dubord, *The Happy Hookers* (privately published, 1988), available at the QCHSM.

CHAPTER 3: SAFETY AND SIGHT

1 Gould referred to the audio compositions as "contrapuntal radio." Glenn Gould, *The Idea of North; The Latecomers; Quiet in the Land,* CBC Records PSCD 2003-3 (CD).

2 Kai Erickson, "Radiation's Lingering Dread," *Bulletin of the Atomic Scientists* 47 (March 1991): 34-39. See the commentary by J. Samuel Walker, *Permissible Dose: A History of Radiation Protection in the Twentieth Century* (Berkeley: University of California Press, 2000), 135-40.

3 Interview of Jim Bayes, Port Elgin, ON, 6 October 2003; interview of Bryan Patterson, Dipper Harbour, NB, 5 August 2003.

4 Françoise Zonabend, *The Nuclear Peninsula,* trans. J.A. Underwood, (Cambridge: Cambridge University Press, 1993 [1989]); Monica Schoch-Spana, "Reactor Control and Environmental Management: A Cultural Account of Agency in the US Nuclear Weapons Complex" (PhD diss., Johns Hopkins University, 1998).

5 Gabrielle Hecht, *The Radiance of France: Nuclear Power and National Identity after World War II* (Cambridge, MA: MIT Press, 1998); Itty Abraham, *The Making of the Indian Atomic Bomb* (London: Zed Books, 1998).

6 Interview of Lorne McConnell, Etobicoke, ON, 25 October 2002; interview of John Wilkinson, Hastings, ON, 17 October 2002; interview of Robert Wilson, Scarborough, ON, 1 November 2001 (for the "shepherd" reference); interview of Frank Caiger-Watson, Kincardine, ON, 3 July 2002; interview of Jan Burnham, St Andrew's, NB, 22 October 2002; interview of Dave McKee, Kincardine, ON, 2 July 2002; interview of Charles Mann, Kincardine, ON, 29 October 2001; and interview of Gerald Black, Grand Bay-Westfield, NB, 28 October 2003. On the spectrum of hazards at defence sites, see Peter Hales, *Atomic Space: Living on the Manhattan Projects* (Urbana: University of Illinois, 1997), 284-89.

7 Interviews of John Wilkinson, Frank Caiger-Watson, Jim Bayes, and Dave McKee; interview of Ken Hill, Quispamsis, NB, 28 October 2003; interview of John Wilkinson (on the temptations to proceed based on trade knowledge).

8 Robert Bothwell, *Nucleus: The History of Atomic Energy of Canada Limited* (Toronto: University of Toronto Press, 1988), 154-66; interview of Charles Mann.

9 Variant versions of the event were told to me by Dave McKee and Charles Mann. It became a frequent "teachable moment" in radiation protection training in later years. There is another version in the University of Toronto Archives (UTA), Bothwell Family Papers, interview of Dr. Gordon Stewart by Robert Bothwell, Deep River, ON, 15 April 1986.

10 Interviews of John Wilkinson and Frank Caiger-Watson.

11 Interview of Ken Elston, Kincardine, ON, 10 October 2002; interviews of John Wilkinson, Lorne McConnell, and Dave McKee.

12 Interviews of Lorne McConnell and John Wilkinson (on deciding how to do his job); interview of Dick Joyce, Underwood, ON, 16 October 2001.

13 Interviews of Lorne McConnell and Ken Elston.

14 Interviews of Ken Elston, John Wilkinson, Lorne McConnell, and Dave McKee.

15 Interviews of John Wilkinson, Frank Caiger-Watson, Robert Wilson, and Lorne McConnell. Shoshona Zuboff describes a similar transition in the computerization of manufacturing processes in *In the Age of the Smart Machine* (New York: Basic Books, 1988), 59, 77. The recent literature on knowledge management is also helpful in untangling the issues the drafting team faced. See David J. Teece, "Knowledge and Competence as Strategic Agents," and Bo Newman, "Agents, Artifacts and Transformations: The Foundations of Knowledge Flows," in Clyde W. Holsapple, ed., *Handbook on Knowledge Management I* (Berlin: Springer, 2003), 138 and 305, respectively; Claus Otto Scharmer, "Self-Transcending Knowledge," in Ikujiro Nonaka and David Teece, eds., *Managing Industrial Knowledge* (London: Sage, 2001), 83-88. Fredrik Ericsson and Anders Avdic formalize a similar interaction in "Information Technology and Knowledge Acquisition in Manufacturing Companies: A Scandinavian Perspective," in Elayne Coakes, Dianne Willis, and Steve Clarke, eds., *Knowledge Management in the Socio-Technical World* (Berlin: Springer, 2002), 121-35.

16 Interview of Bob Ivings, Southampton, ON, 9 October 2003; interviews of Jim Bayes, Dave McKee, and Ken Hill.

17 Ontario Hydro had been a co-sponsor of the Chalk River site since 1955. See Bothwell, *Nucleus*, 206-7.

18 Interviews of Dave McKee, Frank Caiger-Watson, Lorne McConnell, Ken Elston, Ken Hill, Jim Bayes, and Bob Ivings; Thomas Csordas, "Somatic Modes of Attention," *Cultural Anthropology* 8 (1993): 135-56.

19 Interviews of John Wilkinson and Jan Burnham.

20 Schoch-Spana, "Reactor Control," 271-72. The technological differences between Canadian and American reactors would not have been in the foreground in this case. Savannah River was a US heavy water site with radiological hazards similar to those in CANDU plants.

21 Interviews of Robert Wilson and Dick Joyce.

22 Rodney P. Carlisle, "Probabilistic Risk Assessment in Nuclear Reactor," *Technology and Culture* (1997): 922; UTA, Bothwell Family Papers, interview of Lorne McConnell by Robert Bothwell, Toronto, 17 September 1986; Bothwell, *Nucleus*, 149; Schoch-Spana, "Reactor Control," 166; interviews of Jan Burnham and Jim Bayes.

23 Walker, *Permissible Dose*, 104-5.

24 Interview of Dave McKee.

25 Robert Gillette, "'Transient' Nuclear Workers: A Special Case for Standards," *Science* 186, 11 October 1974, 125-29; interviews of Frank Caiger-Watson, Gerald Black, and Stephen Frost, Johnson Settlement, NB, 11 August 2003.

26 UTA, Bothwell Family Papers, Interview of Lorne McConnell by Robert Bothwell; interview of Jan Burnham.

27 John R. Childress and Victoria Broadhead Briant, "Risk Management through Concurrency: A New Work Culture for Improving Safety and Performance," in Ronald A. Knief, *Risk Management: Expanding Horizons in Nuclear Power and Other Industries* (New York: Hemisphere, 1991), 85-89.

28 Nuclear Energy Agency, OECD, *Work Management in the Nuclear Power Industry* (Paris: OECD, 1997), 11-15.

29 Zonabend, *Nuclear Peninsula*, 76-81, 96-97, 101.

30 Interviews of Jim Bayes, Robert Wilson, and Bob Ivings; UTA, Bothwell Family Papers, interview of Lorne McConnell by Robert Bothwell; R. Wilson and D.A. Lee, "Radiation Protection – the Future," 27th Annual Conference of the Canadian Nuclear Association, Saint John, NB, 14-17 June 1987, *Proceedings,* 379-83.

31 Interviews of Jim Bayes, Bob Ivings, and Frank Caiger-Watson; interview of Dave Meneer, Dipper Harbour, NB, 5 August 2003; H.K. Rae, "Heat Transport System," in Atomic Energy Canada Limited, *Canada Enters the Nuclear Age: A Technical History of Atomic Energy of Canada Limited* (Montreal: McGill-Queen's University Press, 1997), 283-87; R. Wilson, G.A. Vivian, W.J. Chase, G. Armitage, and L.J. Sennema, "Occupational Dose Reduction Experience in Ontario Hydro Nuclear Power Stations," *Nuclear Technology* 72 (March 1986): 243-44.

32 Bothwell, *Nucleus,* 296-97; UTA, Bothwell Family Papers, interview of Lorne McConnell by Robert Bothwell; interviews of Jim Bayes, Bob Ivings, and Frank Caiger-Watson; Wilson et al., "Occupational Dose Reduction," 231.

33 Interviews of Dave McKee, Jim Bayes, Bob Ivings, Jan Burnham, and Ken Hill.

34 Zonabend, *Nuclear Peninsula,* 105-13; Schoch-Spana, "Reactor Control," 337.

35 Catherine Bush, *The Rules of Engagement* (Toronto: HarperCollins, 2000), 176.

36 Interviews of Robert Wilson, John Wilkinson, Jim Bayes, Ken Hill, and Dave Meneer; Zonabend, *Nuclear Peninsula,* chap. 5.

37 Private Collection of Dick Joyce, Underwood, ON, Examinations, Radiation Protection, Douglas Point GS, 10 February 1966, 17 March 1966, 6 April 1966. These exams were set by Robert Wilson. See Ontario Hydro, *Radiation Protection Regulations,* 1962. See also interviews of Robert Wilson and John Wilkinson.

38 An incident from this 1964 arbitration before H. Carl Goldenberg, OBE QC, described in the bargaining notes of Ontario Hydro representatives, is retold by union members to emphasize the exceptional knowledge required of nuclear workers. Bob Abbott, Hydro's manager of labour relations, contended that a mechanical maintainer at a nuclear station didn't need to know any more than a mechanical maintainer at a hydraulic site (i.e., a dam). Bob Ivings, a veteran of the Hearn steam plant and the NPD and, in 1964, an operator at Douglas Point, pointing to a metre-high stack of manuals he had piled on a nearby chair, exclaimed, "Bullshit." The lawyer for the union, David Lewis, later a leader of the New Democratic Party, is said to have rephrased: "I am informed by my young friend here that that isn't true," prompting Goldenberg to reply, "Yes, I hear, and I don't want to hear it again." See Ontario Hydro Archives (OHA), Labour Relations files for 1964, submission to Goldenberg regarding the NPD and the DP, with accompanying

notes; Power Workers Union files, "In the matter of an arbitration between the Hydro-Electric Power Commission of Ontario and the Ontario Hydro Employees Union, Local 1000, award," 18 March 1964; interviews of Dave McKee and Bob Ivings.

39 OHA, Labour Relations, 11 February 1964; interviews of Ken Elston, Frank Caiger-Watson, and Robert Wilson. Ontario Hydro specified that its recruits would have Grade 13 and that they would be graduates of the five-year academic high school stream (or equivalent); however, given the composition of the local labour pool, equivalence often must have been offered and accepted.

40 Interview of Jan Burnham, Dan McCaskill, Maces Bay, New Brunswick, 27 October 2003; interview of Stephen Frost.

41 Interview of Robert Wilson.

42 Interviews of Dan McCaskill, Stephen Frost, Dave Meneer, and Jan Burnham; Jan Burnham, *Radiation Protection Training Course (RP)* (New Brunswick Power, 1979, 1981, 1986, 1992, 2002).

43 The quip first appears in Burnham, *RP* (1979), RPT (A)-8.1, 2 and persists (1986), 310; (1992), 308; (2001), 227.

44 Interviews of Dan McCaskill, Ken Hill, John Wilkinson, and Frank Caiger-Watson; Zuboff, *In the Age*, 72-73; Zonabend, *Nuclear Peninsula*, 87, 117.

45 Interviews of Ken Hill, Stephen Frost, Dave Meneer, Jan Burnham, Charles Mann, and Frank Caiger-Watson. Gerald Black's speciality at Point Lepreau was instrumentation, and his interview on this topic is especially valuable. On "affordances" see Zuboff, *In the Age*, 187. In the first four editions of the Burnham manual, Chapter 5 concerns instrumentation. Instrumentation is covered in Burnham, *RP* (1979), RPT (A)-5.4, 16-17; (1986), 201-3; (1992), 202-4; interview of Stephen Frost; Chapter 6 in Burnham (2001). See William H. Halenbeck, *Radiation Protection* (Boca Raton: Lewis, 1994), 126-33.

46 Burnham, *RP* (1979), RPT (A)-5.4, 16-17; (1986), 201-3; (1992), 202-4; interview of Stephen Frost.

47 Burnham, *RP* (1979), RPT (A)-9.6, 4; (1986), 418; (1992), 180, 400; (2001), cover.

48 The "tongs not tongues" play first appears in Burnham, *RP* (1992), 251, and persists in (2001), 188.

49 Burnham gives the instruction straight in *RP* (1981), RPT (A)-11.1, 11, and in (1986), 223. The nipples reference appears in the text in 1992, 293, and is gone by 2001.

50 Burnham, *RP* (1979), RPT (A)-6.1, 10; (1986), 249; (1992), 246; (2001), 184.

51 Interview of Jan Burnham.

52 Burnham, *RP* (1992), 480, 268, 448, 138, 554.

53 Burnham, *RP* (1986), 279-80; (1992), 277.

54 Interviews of Jim Bayes, Bob Ivings, Dan McCaskill, Ken Hill, Stephen Frost, and Barry Schell, Port Elgin, ON, 15 October 2001; OHA, Labour Relations Microfiche Collection, handwritten notes: on bargaining, 2 April 1963, 11 February 1964, 12, or 19 September 1964; on noise, Union's Bargaining Agenda – Nuclear, 29 March 1966; on crotch tears, A.J. Frey, Senior Health Physicist to William McCullough, 1 February 1966.

55 Interviews of Bob Ivings, Dan McCaskill, and Stephen Frost; Schoch-Spana, "Reactor Control," 318-22; Nancy Munn, "Excluded Spaces: The Figure in the Australian Aboriginal Landscape," *Critical Inquiry* 22 (1996): 446-65; Zonabend, *Nuclear Peninsula*, 75 ff.

56 Burnham, *RP* (1979), RPT (A)-9.4, 1-4; a blue and green coded zoning map appears first in 1986, (1986), 400-1, and persists (1992), 398-400.

57 Interviews of Gerald Black, Dan McCaskill, Ken Hill, Jim Bayes, and Bryan Patterson; Schoch-Spana, "Reactor Control," 342.

58 Burnham, *RP* (1992), 286-89; interviews of Ken Hill, Gerald Black, Dan McCaskill, Jan Burham, and Stephen Frost.

59 Burnham, *RP* (1979), RPT (A)-9.4, 2-4, RPT (A)-9.5, 1-7; (1986), 402-14; (1992), 395-410; interview of Lorne McCaskill.

60 Zonabend, *Nuclear Peninsula*, 102-3, 114-5, and note 30 above.

61 Interviews of Robert Wilson, Bob Ivings, Jim Bayes, Ken Hill, Dave Meneer, and Lorne McCaskill; Burnham, *RP* (1992), 460, 466.

Chapter 4: Movement and Sound

1 For an elaborated assessment of the Iroquois new media project within the context of other media forms, see Joy Parr, Jon van der Veen, and Jessica Van Horssen, "The Practicing of History Shared across Differences: Needs, Technologies and Ways of Knowing in the Megaprojects New Media Project," *Journal of Canadian Studies* 43, 1 (2009): 48-58.

2 Kenneth Candy, "Address at the University of Toronto School of Architecture on the Subject of the St. Lawrence Seaway," 15 February 1961; the population of Iroquois was 1,086 in 1951, 1,097 in 1901. See *Census of Canada*, 1951, vol. 1, table 6.

3 "Huge House Moving Job Is Nearing Completion," *Standard and Freeholder*, 16 May 1957, 13. The newspaper account says that 150 houses were moved. Clive and Frances Marin, in their careful history of the area, say that 151 were moved. See Marin and Marin, *Stormont, Dundas and Glengarry, 1945-78* (Belleville: Mika, 1982), 34.

4 See Figure 4.2, a photograph of Cooke Sisty's father and his cart in the dappled shade of old Iroquois.

5 Interview of Rose Sisty, Iroquois, 14 April 2000.

6 Marcel Mauss, "The Notion of Body Techniques," *Sociology and Psychology*, trans. Ben Brewster (London: Routledge and Kegan Paul, 1979), 103-5.

7 Interview of Ron Fader, Iroquois, 10 April 2000. The maps at http://megaprojects.uwo.ca/Iroquois give a good sense of these boundaries and dimensions.

8 Joseph Amato, *On Foot: A History of Walking* (New York: New York University Press, 2004), 16; interview of Jean Shaver, Iroquois, 31 March 2000.

9 David Harvey, *The Condition of Postmodernity: An Inquiry Into the Origins of Cultural Change* (Oxford: Blackwell, 1989), 240.

10 Amato, *On Foot*, 276.

11 Mimi Sheller, "Automotive Emotions," *Theory, Culture and Society* 21, 4/5 (2004): 226.

12 Ibid.; Michel de Certeau, "Walking in the City," in *The Practice of Everyday Life* (Berkeley: University of California Press, 1984), 97-110; "Iroquois Residents Face Future with Confidence," *Standard and Freeholder*, 26 June 1954, 16.

13 Paul C. Adams, "Peripatetic Imagery and the Peripatetic Sense of Place," in Paul C. Adams, Steven Hoelscher, and Karen E. Till, eds., *Textures of Place: Exploring Humanist Geographies* (Minneapolis: University of Minnesota Press, 2001), 188-89; Nigel Thrift, "Driving in the City," *Theory, Culture and Society* 21, 4/5 (2004): 42.

14 Lewis Mumford, "The Highway and the City," in *The Highway and the City* (New York: New American Library, 1964), 237 and 243. The essay first appeared in the *Architectural Record* 123 (April 1958): 179-86.

15 Amato, *On Foot*, 156.

16 Interview of Ron Fader.

17 Follow the walks and listen to the villagers' commentaries at http://megaprojects.uwo.ca/ Iroquois. Interview of Jean Shaver, Iroquois, 31 March 2000; Keith Thomas, "Introduction," 7; Peter Burke, "The Language of Gesture in Early Modern Italy," 72, 77; Maria Bogucka, "Gesture, Ritual and Social Order in Poland," 206; and Herman Roodenburg, "The 'Hand of Friendship,'" 159. All of the preceding appear in Jan Bremmer and Herman Roodenburg, *A Cultural History of Gesture* (Cambridge: Polity, 1991). See also Rebecca Solnit, *Wanderlust: A History of Walking* (New York: Viking, 2000), 9, 10; Amato, *On Foot*, 217.

18 Interview of Joyce Fader, Iroquois, 17 March 2000.

19 Interview of Keith Beaupre, Iroquois, 10 April 2000.

20 Interview of Erma Stover, Iroquois, 5 April 2000.

21 Interview of Joseph Roberts, Iroquois, 28 March 2000.

22 de Certeau, *The Practice*, 99.

23 Solnit, *Wanderlust*, 5, 178.

24 Mumford, "The Highway," 244.

25 Interview of Ambert Brown, Iroquois, 21 March 2000.

26 Interview of Joseph Roberts. There are images of the powerhouse at http://megaprojects. uwo.ca/Iroquois.

27 Interview of Keith Beaupre, Iroquois, 10 April 2000; interview of Hilda Banford, Iroquois, 13 April 2000.

28 Interview of Joseph Roberts. There is a walk both along the shore and along the tracks at http://megaprojects.uwo.ca/Iroquois.

29 Interview of Joseph Roberts.

30 Interview of Jean Shaver.

31 Interview of Ambert Brown.

32 Interview of Jean Shaver.

33 Interviews of Keith Beaupre and Joseph Roberts.

34 Interviews of Hilda Banford, Ambert Brown, and Erma Stover. There are images of these places on the front walk to the west of the commercial district on http://megaprojects. uwo.ca/Iroquois.

35 "Soon to Be Bisected by Seaway, Point Iroquois an Apple Centre," *Standard and Freeholder* (Cornwall), 3 November 1954, 3, 18; "Iroquois Is Humming with Activity as Seaway Works Are Accelerated," *Standard and Freeholder,* 18 June 1955; "Seaway a Poser for Iroquois," *Standard and Freeholder,* 17 May 1952 (picture of Miss Sally Tindale looking through her orchard to the river and the site of the control dam); "Annual Picnic MacIntosh Clan Largely Attended," *Standard and Freeholder,* 9 September 1954. Interview of Jack Fetterly, Iroquois, 20 March 2000; interview of Carl Van Camp, Iroquois, 4 April 2000; interview of Caroline Grisdale Robertson, Iroquois, 30 March 2000; interview of Joseph Roberts; interview of Leo Merkeley, Iroquois, 2 April 2000. See also Maida Parlow French, *Apples Don't Just Grow* (Toronto: McClelland and Stewart, 1954). The best way to make sense of the remade Point is to look at the Seaway Authority engineering plan accessible along the top of the walking tour at http://megaprojects.uwo.ca/Iroquois. The air photos of the river by the Point show the rapids that sounded there.

36 Interview of Don Thompson, Iroquois, 31 March 2000; interview of Anne Thompson, Iroquois, 29 March 2000; interview of Joseph Roberts; interview of Joyce Fader.

37 Sherban Cantacuzino, *Wells Coates: A Monograph* (London: G. Frazer, 1978); Laura Cohn, *The Door to a Secret Room: A Portrait of Wells Coates* (Aldershot: Scolar, 1999); Wells Coates

Exhibition Committee, *Wells Coates: Architect and Designer* (Oxford: Oxford Polytechnic Press, 1978); Iroquois Civic Centre, Council minute books, 7 August 1952, 14 September 1954.

38 Iroquois Civic Centre, Map, "Land Use, Iroquois New Town"; "General Land-Use Plan," 19 February 1953; "Saunders States Site in Marine Clay Area," *Standard and Freeholder,* 14 October 1954.

39 "Iroquois, Matilda Twp. Differ on Rehab Plans," *Standard and Freeholder,* 15 September 1954, 3; "Officials Blast Hydro Rehab Plans," *Standard and Freeholder,* 21 August 1954; "Planner Shocked by Iroquois Move," *Globe and Mail* (Toronto), 8 November 1954; "Iroquois Move Stuns Advisor from London," *Ottawa Citizen,* 8 November 1954.

40 "Hydro Plan Disappoints Iroquois," *Standard and Freeholder,* 18 August 1954.

41 "Saunders Allays Citizens' Fears on Rehabilitation," *Standard and Freeholder,* 25 August 1954.

42 Janet Davis Scrapbooks, "Hydro Plan Has Possibilities Reeve L.C. Davis Informs Villagers,' *Iroquois Post,* 21 October 1954.

43 "Coates Says Hydro Plan 'Inexcusable,'" *Standard and Freeholder,* 21 August 1954.

44 Nancy Martin, "National Triumph and Personal Tragedy: Towards a More Complete History of the Creation of the St. Lawrence Seaway and Power Project, 1954 to 1959" (Honours paper, Queen's University, 1994), 18.

45 Interviews of Leo Merkley, Ron Fader, and Janet Davis, Iroquois, 29 March 2000.

46 Interviews of Ron Fader, Ambert Brown, and Shirley Fisher, Iroquois, 4 April 2000.

47 R. Murray Schafer, *The Tuning of the World* (New York: Knopf, 1977), 9.

48 Interview of Shirley Kirkby Carnegie, Iroquois, 17 March 2000; interviews of Don Thompson, Anne Thompson, and Les Cruickshank, Iroquois, 7 April 2000.

49 Interviews of Carl Van Camp and Fred Brouse, Iroquois, 6 April 2000; interview of Don Thompson; interview of Leo Merkley.

50 Interviews of Carl Van Camp, Fred Brouse, Don Thompson, and Leo Merkley.

51 Bruno Latour, *We Have Never Been Modern* (Cambridge, MA: Harvard University Press, 1993), 104.

52 Interviews of Leo Merkley, Ron Fader, Janet Davis, and Rose Sisty.

53 Interviews of Carl Van Camp, Shirley Fisher, and Jean Shaver.

54 Interviews of Les Cruickshank, Joyce Fader, Jean Shaver, Leo Merkeley, Ron Fader, Fred Brouse, Ambert Brown, and Gwyneth Casselman, Iroquois, 12 April 2000; interviews of Shirley Kirkby-Carnegie, Caroline Grisdale Robinson, and Jack Fetterly, 20 March 2000. "A Changed Iroquois Point" (photo), *Standard and Freeholder,* 12.

55 Interviews of Carl Van Camp, Hilda Banford, Ron Fader, Ambert Brown, and Jack Fetterly; "Cutting Down Trees Termed 'Squeeze Play,'" *Standard and Freeholder,* 25 January 1956, 3; "Trees Tumble in Entire Headpond Area," *Standard and Freeholder,* 21 March 1957, 13; "Seaway Valley Residents Discover Pressure Can Alter Hydro's Plans," *Globe and Mail* (Toronto), 23 December 1955, 4.

56 The whole expropriation set is at the Upper Canada Village Archives in Morrisburg, where the digital copies we created for Iroquois are available for examination.

57 Interview of Les Cruickshank.

58 Interview of George Hickey, Ingleside, 6 April 2000; Martin, "National Triumph," 32-33, concerning Donald Stuart.

59 Interview of Ron Fader; "Headpond Clearance Nears End," *Standard and Freeholder,* 21 March 1958, 9, 10, on the barren look of the shoreline.

60 Interviews of Shirley Kirkby-Carnegie, Joyce Fader, Janet Davis, Caroline Grisdale Robertson, Shirley Fisher, and Ambert Brown; Ken Kirkby, "New Iroquois Townsite Undergoes Big Change," *Standard and Freeholder*, 13 January 1956.

61 Interviews of Ron Fader, Shirley Kirkby-Carnegie, and Shirley Fisher.

62 T.L. Hills, *The St. Lawrence Seaway* (London: Methuen, 1959), 99–117; B. Paul Wisnicki, "The St. Lawrence Seaway and Power Developments," *BC Professional Engineer*, February 1957, 13–17; "US Writer Takes Look at Seaway District," *Standard and Freeholder*, 16 June 1954, 3, 22; Marin and Marin, *Stormont*, 21; "Rapids Stilled after Epic Struggle," *Standard and Freeholder*, 4 April 1957, 1; "St. Lawrence through Many Eyes Is Pictured in Fine Anthology," *Standard and Freeholder*, 16 August 1958, a review of Jean Gogo, *Lights on the St. Lawrence: An Anthology* (Toronto: Gage, 1958); "Silencing of Rapids Tough Job," *Standard and Freeholder*, 30 March 1957, 13, 14.

63 "Hydro Vice-Chairman Foresees St. Lawrence Valley Becoming Another Ruhr Area," *Standard and Freeholder*, 11 October 1956, 13, 14; "Planning Department Aids In Rehabilitation Program," *Standard and Freeholder*, 15 February 1958, 7.

64 Interviews of Jack Fetterly and Glen Cunningham, Iroquois, 5 April 2000.

65 Hills, *St. Lawrence Seaway*, 9, 50, 77–84; Norman D. Wilson, "The Rehabilitation of the St. Lawrence Communities," for the Advisory Committee on Reconstruction, April 1943, sec. 4, 19–24; J.H. Jackson, "The St. Lawrence Power Project Rehabilitation: A Review of Major Factors," *Engineering Journal* 75 (February 1960): 67.

66 Interview of Ann Thompson.

67 Interview of Janet Davis, 29 March 2000; Clark Davey, "Seaway Valley Residents Satisfied with Hydro's Efforts in Moving Homes to New Sites," *Globe and Mail* (Toronto), 11 July 1956, 9.

68 Carleton Mabee, *The Seaway Story* (New York: Macmillan, 1961), citing General A.G.L. McNaughton, in his capacity as a commissioner of the International Joint Commission 174, 214.

69 Sara Bowser, "The Planners' Part," *Canadian Architect*: February 1958, 3–9; and Peter Stokes, "St. Lawrence: A Criticism," ibid., 46.

70 "Seaway Valley Transformation," *Standard and Freeholder*, 21 March 1958, 4; Davey, "Seaway Valley Residents," 9; Chris Jermyn, "Some St. Lawrence Seaway Communities, 1959–1969," *Canadian Geographical Journal* 79 (1969): 158, 163.

71 "Plan New Highway across Control Dam Linking NY, Ontario," *Standard and Freeholder*, 17 August 1954, 1, 3.

72 Maybee, *Seaway Story*, 211; Marin and Marin, *Stormont*, 28–29; "New Homes Find Little Favor," *Standard and Freeholder*, 20 August 1954. Interview of George Hickey, Ingleside, 6 April 2000; interviews of Janet Davis and Rose Sisty. Ontario Hydro, "Proposals for Rehabilitation of Communities Affected by St. Lawrence Power Project," August 1954, including elevations of the house plans and street maps of new Iroquois, Morrisburg, and New Towns nos. 1 and 2, later named Ingleside and Long Sault, respectively.

73 "Compensation Key to Rehab Problems," *Standard and Freeholder*, 9 October 1954, 6; "Plan New Highway across Control Dam Linking NY, Ontario," *Standard and Freeholder*, 17 August 1954, 3; Marin and Marin, *Stormont*, 28–29, 36–40.

74 Interviews of Ann Thompson, Don Thompson, Ambert Brown, Fred Brouse, Shirley Fisher, Shirley Kirkby-Carnegie, Jack Fetterly, and Ann Davis; Clark Davey, "Hydro Pay at Iroquois Benefits Seaway Valley," *Globe and Mail*, 11 November 1955, 13.

75 Solnit, *Wanderlust*, 257; Martin, "National Triumph," 50, citing George Hickey; "Plan New Highway across Control Dam Linking NY, Ontario," *Standard and Freeholder*, 17 August 1954; "New Highway?" *Standard and Freeholder*, 17 August 1954, 1, 3; "New Towns Becoming 'Live' Communities," *Standard and Freeholder*, special Seaway-Power Report and Tourist Guide, 29 June 1957, 14.

76 Stokes, "St. Lawrence: A Criticism," 44.

77 Yi-fu Tuan, *Topophilia: A Study of Environmental Perception, Attitudes and Values* (Englewood Cliffs: Prentice-Hall, 1974), 175.

78 Solnit, *Wanderlust*, 249.

79 "New Shopping Centre Opening On May 11," *Standard and Freeholder*, 4 May 1957, 13.

80 Interviews of Jack Fetterly, Don Thompson, and Rose Sisty.

81 Interviews of Gwyneth Casselman, Les Cruickshank, and Erma Stover.

82 Interviews of Jack Fetterly, whose garage was across the highway in new Iroquois; Joyce Fader, who, in the new town, worked with her husband in a furniture store; Erma Stover, whose parents continued their grocery business in the new town. "Hard to Find Place to Sit in Morrisburg," *Standard and Freeholder*, 15 March 1958, 9.

83 Marin and Marin, *Stormont*, 41, 48; interviews of Keith Beaupre (who continued to operate his jewellery store in the new town), Shirley Kirkby-Carnegie, Hilda Banford, Erma Stover, Gwyneth Casselman, Fred Brouse, and Rose Sisty.

84 Interviews of Ron Fader, Fred Brouse, Les Cruickshank, and Keith Beaupre.

85 Interview of Rose Sisty.

86 "Transition Period Near End in Iroquois," *Standard and Freeholder*, 23 April 1957, 4; "Heavy Safe Hauled Out Store Window: Two Trucks Stolen," *Standard and Freeholder*, 3 June 1957, 9; "Hydro Will Supply Two Extra Police at Iroquois Plaza," *Standard and Freeholder*, 7 June 1957, 13; "Robberies Arouse Iroquois Merchants," *Standard and Freeholder*, 7 August 1957, 13; "$500 Stolen in Robbery At Iroquois," *Standard and Freeholder*, 14 August 1957, 13; "Thieves Strike Again, $2400 in Loot Stolen," *Standard and Freeholder*, 24 February 1958, 9; Jane Jacobs, *The Death and Life of Great American Cities* (New York: Vintage, 1961), chap. 2.

87 Interview of Les Cruickshank.

88 "Hydro Plan for Iroquois Waterfront Fails to Please," *Standard and Freeholder*, 23 March 1956, 2; "Waterfront an Issue at Iroquois," *Standard and Freeholder*, 10 July 1956, 11; interview of Fred Brouse; "Park Plan Is Outlined to Iroquois Ratepayers," *Standard and Freeholder*, 19 October 1956, 9. Similarly, from Thorold Lane, reeve of Osnabruck Township and chairman of the Joint Councils of the Affected Municipalities, "A Submission on the Development and Improvement of the Shoreline and Flooded Areas Created by the St. Lawrence Seaway and Power Project," 14 August 1956, 1, 2, of Exhibit A, Upper Canada Village Archives.

89 Interviews of Carl Van Camp, Fred Brouse, Jack Fetterly, and Lee Shaver (who worked as a heavy equipment operator on the site). See Ronald E. Richard, Walter G. Rooke, and George H. McNevin, *Developing Water Resources: The St. Lawrence Seaway and the Columbia/Peace Power Projects* (Toronto: Ryerson and MacLean-Hunter, 1969), 30-34; "Water Level Issue Causing Major Concern," *Iroquois Post*, 28 July 1955; "No Decision Yet on Iroquois Park, Seek Information," *Recorder and Times* (Brockville), 7 May 1956.

90 "Plan of the Rehabilitated Village of Iroquois When Park Land Is Completed," *Standard and Freeholder*, 5 October 1957, 22; "Iroquois Park Plans Are Unveiled by Hydro," 11 October 1956, 13.

91 These quotes are from two letters printed side by side, *Standard and Freeholder,* 12 June 1954, 6. The first is written by "One who will be moved" and the second is written by C.S. Pattenden of Dickinson's Landing in Osnabruk Township.

92 "New Business Centre Lighting Plans Laid," *Standard and Freeholder,* 15 September 1956, 9, citing J.H. Jackson, Hydro engineer, at a meeting with Iroquois village council.

93 Donald Worster, "Thinking Like a River," in *The Wealth of Nations* (New York: Oxford University Press, 1993), 124-25, and, in the same volume, "The Transformation of the Earth," 48.

94 Proportion of residents by age group (%), 1951

	0-4	5-9	10-14	15-19	20-24	25-34	35-44	45-54	55-65	65-69	>70
Iroquois	10	9.3	6.4	8.1	7.8	13.0	10.9	8.6	11.0	3.7	10.7
Ontario	11	8.7	7.0	6.8	7.6	16.0	13.9	11.2	8.5	3.3	5.3
Iroquois/ Ontario	.9	1.06	.91	1.19	1.02	.81	.78	.77	1.3	1.12	2.01

Census of Canada, 1951, vol. 1, table 23.

95 Interviews of Fred Brouse, Rose Sisty, Shirley Kirkby-Carnegie.

96 "Saving the Topsoil of the Seaway," and cartoon, "Looks Good to Him," reprinted from the *Sudbury Star* in *Standard and Freeholder,* 25 July 1955, 6. Similarly, "Jest around Cornwall," a cartoon featuring a local directing a tourist, with a station wagon and a garden fork, towards "rhubarb roots, lilacs, raspberry canes, shrubs, topsoil, perennials, sods, topsoil" along a road lined with remnant house foundations leading towards the superstructure of a Hydro dam. See *Standard and Freeholder,* 10 May 1958, 13. See also "Inundation Day Quotes," *Standard and Freeholder,* 28 June 1958, 3.

97 University of Toronto, School of Social Work, *Round Table on Man and Industry, Community Survey Report, St. Lawrence Impact Area,* vol. 4, part 2, Town of Iroquois, 40, 39, 41. Near Gagetown, couples resettled from the Baselands, despite having been dispersed, and maintained the dance and musical traditions of the Lawfield Road, gathering in the No. 4 LOL Hall, which was moved from Dunn's Corner to Queenstown.

98 University of Toronto, *Community Survey Report,* 29, 41.

99 Interview of Carl Van Camp.

100 Interview of Joyce Fader. All the following quotations are from this interview. Joyce Fader became the historian of the village. In mid-life she suffered from a disabling accident, recovered, and retrained as a hairdresser. Her work to make new Iroquois a good place to live has been prodigious, and her insights about the effects of the displacement on her neighbours cut to the heart of their shared predicament.

CHAPTER 5: TIME AND SCALE

1 For a description of Van Horssen's experiences with Val Morton in researching for this site, see Joy Parr, Jon van der Veen, and Jessica Van Horssen, "The Practicing of History Shared across Differences: Needs, Technologies and Ways of Knowing in the Megaprojects New Media Project" *Journal of Canadian Studies* 43, 1 (2009): 42-48.

2 Adam Smith, *The Wealth of Nations* (Oxford: Clarendon, 1869), book 1, chap. 4, 29; David Ricardo, *The Principles of Political Economy and Taxation* (London: Dent, 1911), chap. 1, sec. 1, 5-11. A contemporary analysis of these differences with close resonances to the case

considered here is found in Thomas Michael Power, *Lost Landscapes and Failed Economies* (Washington, DC: Island Press, 1996), 19-28.

3 There is a film clip of the *Minto*'s fiery demise on http://megaprojects.uwo.ca/ArrowLakes.

4 *The Reckoning* (Vancouver: CBC, 1974). Of late, students are as likely to be reminded of the 2003 movie *Northfork*, which was set in Montana in 1955 as townspeople prepared to leave before the rising waters of a reservoir.

5 "Byline Burpy," *Arrow Lakes News*, 19 February 1975, 2.

6 Juliet B. Schor, *The Overworked American: The Unexpected Decline of Leisure* (New York: Basic Books, 1991), chap. 5.

7 Paula Pryce, *"Keeping the Lakes' Way": Reburial and the Re-creation of a Moral World among an Invisible People* (Toronto: University of Toronto Press, 1999), 7-18, 31-55, 154; Eileen Delehanty Pearkes, *The Geography of Memory: Recovering Stories of a Landscape's First People* (Nelson: Kutenai House Press, 2002), 26-29.

8 Harry Holbert Turney-High, *Ethnography of the Kutenai* (Menasha, WI: Memoirs of the American Anthropological Association No. 56, 1941); Pryke, *"Keeping the Lakes' Way,"* 29; Randy Bouchard and Dorothy Kennedy, *First Nations Ethnography and Ethnohistory in British Columbia's Lower Kootenay/Columbia Region* (Castlegar, BC: Columbia Power Corporation, August 2000 [April 2005]), 61, 76, 82, 85, 88, 91.

9 Pearkes, *Geography of Memory*, 21; Pryce, *"Keeping the Lakes' Way,"* is a compelling account of subsequent Sinixt struggles to honour their ancestral obligations north of the forty-ninth parallel; Bouchard and Kennedy, *First Nations Ethnography*, 100-1. On the Colvilles and the changing Columbia in the twentieth century, see Blaine Harden, *A River Lost: The Life and Death of the Columbia* (New York: Norton, 1996), particularly chap. 6, 100-16.

10 Matthew D. Evenden, *Fish versus Power: An Environmental History of the Fraser River* (Cambridge: Cambridge University Press, 2004), 2. See Chris Spicer's testimony concerning the Columbia flooding, reprinted in Donald Waterfield, *Land Grab: Oliver Buerge versus the Authority* (Toronto and Vancouver: Clarke, Irwin, 1973), 63.

11 James Scott, *Seeing Like a State: How Certain Schemes to Improve the Human Condition Have Failed* (New Haven: Yale University Press, 1998), presents a fine and varied disquisition on the origins and implementation of this idea in modern times. For an application of Scott's thought to the Arrow Lakes, see Tina Loo, "People in the Way: Modernity, Environment, and Society on the Arrow Lakes," *BC Studies* 142 (Summer/Autumn 2004): 43-77.

12 Rosemarie Parent, "The Story of Sam Henry," Arrow Lakes Historical Society, http://alhs-archives.com/articles/samhenry.html, viewed 28 July 2006; Milton Parent, *Port of Nakusp* (Nakusp: Arrow Lakes Historical Society, 1992), 62, 96, 123, 128, 134-35, 183, 217, 264-65; *Arrow Lakes News*, 27 December 1922, 23 March 1939.

13 Parent, *Port of Nakusp*, 333-35. See, similiarly, Richard Cole Harris, "Industry and the Good Life around Idaho Peak," in *The Resettlement of British Columbia: Essays on Colonialism and Geographical Change* (Vancouver: UBC Press, 1997) 194-218; and Daisy Phillips, *Letters from Windermere, 1912-1914,* ed. R. Cole Harris and Elizabeth Phillips (Vancouver: University of British Columbia Press, 1984).

14 J.W. Wilson, *People in the Way: The Human Aspects of the Columbia River Project* (Toronto: University of Toronto Press, 1973), 8-9; Donald Waterfield, *Continental Waterboy: The Columbia River Controversy* (Toronto: Clarke, Irwin, 1970), 8-9; Barbara MacPherson, "Columbia River Treaty Committee, Part 1, Arrow Lakes Report," prepared for the Nakusp and District Round Table and Economic Development Board, 25 May 1994, 4.

15 The land Sam Henry, the Buesnels, and the Spicers developed is shown in photographs at http://megaprojects.uwo.ca/ArrowLakes. Jean Spicer's photographs of their market garden and the nearby shoreline are there too.

16 Arrow Lakes Historical Society clipping files, Horace Waterfield, Donald Waterfield, George Brown, Freda Brown Waterfield; interview of Nigel and Ruth Waterfield, Nakusp, 2 April 2004; Parent, *Port of Nakusp*, 247-48.

17 R. Ronson, "Breaking New Farms in the BC Bush," *Family Herald and Weekly Star*, 31 May 1951, 3.

18 R. Ronson, "How I Get Bigger Cream Cheques," *Family Herald and Weekly Star*, 5 October 1950, 13; "How I Planned My Poultry Plant," *Family Herald and Weekly Star*, 18 January 1951, 10; "A Home-Made Brooder – Dutch Oven," *Family Herald and Weekly Star*, 29 March 1951, 16-17; "Select the Right Trees – Shade and Sunshine for Layers," *Family Herald and Weekly Star*, 22 March 1956, 16; "Shearing the Golden Fleece," *Family Herald and Weekly Star*, 24 May 1951, 5, 14.

19 R. Ronson, "Comments on the Economy of Eggs," *Family Herald and Weekly Star*, 8 February 1951, 20; "The Secret of Success in Poultry," *Family Herald and Weekly Star*, 26 April 1951, 17; "The Economy of Eggs: Poultryman Peeks around the Corner," *Family Herald and Weekly Star*, 26 January 1956, 15.

20 R. Ronson, "Protecting Pullets from Pests," *Family Herald and Weekly Star*, 21 September 1950, 20; "Looking after Sheep in the Fall," *Family Herald and Weekly Star*, 19 October 1950, 17.

21 R. Ronson, "Experience Teaches ... Borrowed Roosters Carry Croup," *Family Herald and Weekly Star*, 8 December 1955, 10; "Spartan Apples Show Weakness," *Family Herald and Weekly Star*, 3 April 1952, 17; "Prospecting Goes Modern," *Family Herald and Weekly Star*, 22 April 1954, 30. See also R. Ronson, "The Biogeochemical Method: A Modern Prospecting Technique," *Western Miner*, June 1952, 22, 24.

22 Mihaly Csikszentmihalyi, *Flow: The Psychology of Optimal Experience* (New York: Harper and Row, 1990).

23 R. Ronson, "The Vanishing Woodpile," *Family Herald and Weekly Star*, 29 April 1954, 5.

24 R. Ronson, "Selecting Unprofitable Birds," *Family Herald and Weekly Star*, 7 June 1951, 16, 17; "How to Deal with Lambing Troubles," *Family Herald and Weekly Star*, 9 April 1950, 15; "It's Safer to Feed Horses Good Hay," *Family Herald and Weekly Star*, 12 June 1952, 13.

25 R. Ronson, "'Six Days Shalt Thou Labor,'" *Family Herald and Weekly Star*, 8 November 1951, 5.

26 Joy Parr, *Domestic Goods: The Material, the Moral and the Economic in the Postwar Years* (Toronto: University of Toronto Press, 1999); Joan Sangster, *Dreams of Equality: Women on the Canadian Left, 1920-1950* (Toronto: McClelland and Stewart, 1989); Alvin Finkel, *Our Lives: Canada after 1945* (Toronto: Lorimer, 1997); Magda Fahrni, *Household Politics: Montreal Families and Postwar Reconstruction* (Toronto: University of Toronto Press, 2005).

27 R. Ronson, "Work: Vice or Virtue?" *Family Herald and Weekly Star*, 16 February 1956, 3, 9; and Ibid., "Voice of the Farm," *Family Herald and Weekly Star*, 8 March 1956, 2.

28 R. Ronson, "The Vanishing Mixed Farm," *Family Herald and Weekly Star*, 10 May 1936, 3, 23. Donald Waterfield's other writings on child-rearing are: "Ferdinand Is Dangerous," *Family Herald and Weekly Star*, 21 June 1951, 5; and "The Children Bag Two Bags of Bears," *Family Herald and Weekly Star*, 5 March 1953, 16. See also R. Ronson, "The

Comic Corruption," *National Parent-Teacher,* June 1950, 23-25; and Ibid., "An Alternative to Canned Entertainment," *New Haven Teacher's Journal,* n.d., 16-18.

29 Interview of Janet Spicer, Nakusp, 7 July 2003; "Chamber Opposes High Level Dam, Lakes," *Arrow Lakes News,* 24 December 1959, 1; "Nakusp Chamber Executive Discuss High Arrow Dam," *Arrow Lakes News,* 4 February 1960; "Nakusp WI Asked to Protest High Arrow," *Arrow Lakes News,* 11 February 1960, 1; Letter to the Editor by James Hardwicke, "Useless Destruction," *Arrow Lakes News,* 25 May 1961, 2; Letter to the Editor by Florence Leary, "Asks for Planned Columbia Development," *Arrow Lakes News,* 15 June 1961, 1; "Need Clear Thinking," *Arrow Lakes News,* 6 July 1961, 2; "Why Now a High Arrow?" *Arrow Lakes News,* 24 August 1961, 2; "High Arrow Fight Not Over," *Arrow Lakes News,* 6 February 1961, 2. On these arguments about Hanford, Washington state, and the Columbia dams there, see Richard White, *Organic Machine: The Remaking of the Columbia* (New York: Hill and Wang, 1995), 96.

30 Evenden, *Fish versus Power,* 222, 147.

31 Donald Worster, "Two Faces West: the Development Myth in Canada and the US," in Paul Hirt, *Terra Pacifica: People and Place in the Northwest States and Western Canada* (Pullman: Washington State University Press, 1998), 86-87. See also Donald Worster, *Rivers of Empire: Water, Aridity, and the Growth of the American West* (New York: Oxford University Press, 1992), 260, 265-67, 285, 333; and Harden, *A River Lost,* 100-16. The terms of the Columbia Treaty specified a renegotiation after thirty years, and in 1992 downstream benefits of the dams were reallocated to the Columbia Basin Trust to pay for environmental and social remediation in the region north of the forty-ninth parallel.

32 Neil Swainson, *Conflict over the Columbia: The Canadian Background to an Historic Treaty* (Montreal: McGill-Queen's University Press, 1979); John Krutilla, *The Columbia Basin Treaty: The Economics of an International River Basin* (Baltimore: Johns Hopkins University Press, 1967); Michael Blumm, *Sacrificing the Salmon: A Legal and Policy History of the Decline of Columbia Basin Salmon* (Den Bosch, The Netherlands: BookWorld Publishers, 2002); Susan B. Toller, "Sustainability and Hydro Development in the Columbia River Basin" (MSc thesis, University of British Columbia, 1994); Peter Ommundsen, *Management Guidelines Ungulate Winter Range at Lower Arrow Lake, British Columbia* (Castlegar, BC: Selkirk College, Department of Environmental Studies, 1983); A. Scholz, Kate O'Laughlin, David Geist, Jim Uehara, Dee Peone, Luanna Fields, Todd Kleist, Ines Zozaya, Tim Peone, and Kim Teesaturskie, *Compilation of Information on Salmon and Steelhead Total Run Size, Catch and Hydropower-Related Losses in the Upper Columbia Basin, above Grand Coulee Dam* (Cheney, WA: Fisheries Technical Report 2, Upper Columbia United Tribes Fisheries Center, Eastern Washington University, Department of Biology, 1985); Joseph E. Taylor, *Making Salmon: An Environmental History of the Northwest Fisheries Crisis* (Seattle: University of Washington Press, 1999), 228-29, 246-48.

33 Eric C. Ewart, "Setting the Pacific Northwest Stage," in Hirt, *Terra Pacifica,* 7.

34 Angela K. Martin, "The Practice of Identity and an Irish Sense of Place," *Gender, Place and Culture* 4, 1 (1997): 108.

35 Wilson, *People in the Way,* 169.

36 Michael Polanyi, *The Tacit Dimension* (New York: Anchor Books, 1967), 4.

37 John Gray, "Open Spaces and Dwelling Places: Being at Home on Hill Farms in the Scottish Borders," *American Ethnologist* 26 (1999): 440-60, reprinted in Sethna Low and Denise Lawrence-Zuniga, *The Anthropology of Space and Place* (Oxford: Blackwell, 2003), 225-27; Donald Worster, "Thinking Like a River," in *The Wealth of Nations* (New York: Oxford

University Press, 1993), 124-25, and, in the same volume, "The Transformation of the Earth," 48. See also Wilson, *People in the Way,* 169.

38 These passages are from Wilson's twenty-year follow-up study. See J.W. Wilson and Maureen Conn, "On Uprooting and Rerooting: Reflections on the Columbia River Project," *BC Studies* 58 (1983): 40-54, at 44.

39 Edward Goldsmith and Nicholas Hildyard, *The Social and Environmental Effects of Large Dams* (Camelford, Cornwall: Wadebridge Ecological Centre, 1984), 17.

40 "Slope not in Hydro Proposals," *Arrow Lakes News,* 31 March 1966, 1; interview of Nigel and Ruth Waterfield; Wilson, *People in the Way,* 76.

41 "Dam Looming in Ottawa," *Arrow Lakes News,* 2 February 1961, reprinted from *Ottawa Journal,* 19 January 1961.

42 Patrick McCully, *Silenced Rivers: The Ecology and Politics of Large Dams* (London: Zed Books, 1996), 10.

43 The phrase is from an interview with William Barrow, a tugboat operator long on the lakes. It captures well the transcendent nature with which inhabitants endowed the river they knew. See interview of William Barrow, Nakusp, BC, 11 July 2003.

44 Edward S. Casey, *Remembering: A Phenomenological Study* (Bloomington: Indiana University Press, 1987), 194, elaborated in Steven Feld, "Places Sensed, Senses Placed: Toward a Sensuous Epistemology of Environments," in David Howes, ed., *Empire of the Senses: The Sensual Culture Reader* (Oxford: Berg, 2005), 181-87. See also Edward Casey, *Getting Back into Place: Toward a Renewed Understanding of the Place-World* (Bloomington: Indiana University Press, 1993).

45 "Opposes High Arrow," *Arrow Lakes News,* 24 March 1960, 6.

46 Interview of Winnie Ehl, 9 July 2002; interview of Nigel and Ruth Waterfield.

47 Interview of William Barrow; interview of Helen and Oliver Buerge, Burton, BC, 4 April 2004; interview of Ernie Orr, Nakusp, BC, 8 July 2003. See the pictures of the shoreline from the Morton family collection at http://megaprojects.uwo.ca/ArrowLakes. The bales are canary grass, watered by the rising lake in spring and harvested from the flood plain for fodder. Morton devised a self-feeding storage system for this silage. Jean Spicer's collection includes a photo of her daughters on the emerging beaches of the west side of the lake.

48 Interview of Ernie Roberts, Brouse, Nakusp, BC, 11 July 2003.

49 Interview of Ernie Orr.

50 "Discuss Arrow Lakes Rehabilitation at Nakusp Meet," *Arrow Lakes News,* 19 April 1962, 1; "Dam Reservoir Levels to Be Marked in Area," *Arrow Lakes News,* 31 May 1962, 1; "Stakes Mark Dam Level at Nakusp," *Arrow Lakes News,* 19 July 1962, 1.

51 "Surveyors Stakes Mark Height of Water," *Arrow Lakes News,* 2 August 1962, 1.

52 "High Arrow Dam Not Popular in Arrow Valley," *Arrow Lakes News,* 4 February 1960, 2; "We Live in Unspoiled Habitable Regions," *Arrow Lakes News,* 22 December 1960, 6; "What Joy in a Valley of Mud and Silt," *Arrow Lakes News,* 12 January 1961, 3; "Arrow Lakes People See No Advantage to Compensate Their Great Lifetime-Loss," *Arrow Lakes News,* 19 January 1961, 2.

53 "Byline Burpy," *Arrow Lakes News,* 30 April 1969, 2.

54 Ibid., 25 February 1970, 2. There are pictures of the Spicer garden at draw down, when the reservoir was essentially empty, on http://megaprojects.uwo.ca/ArrowLakes. Photographs by Joy Parr summer 2003.

55 "Changes in Valley Hard to Accept," *Arrow Lakes News,* 21 September 1967, 3. On the burden of the unnatural pattern of flow variations on dammed rivers, see McCully, *Silenced*

Rivers, 46; Mats Dynesius and Christer Nillsson, "Fragmentation and Flow Regulation of River Systems in the Northern Third of the World," *Science*, 4 November 1994, 4.

56 "Byline Burpy," *Arrow Lakes News*, 30 April 1969, 2.

57 Ibid., 7 January 1970; 18 March 1970, 4; photo and caption, 17 July 1974, 1. Interview of Ernie Orr.

58 "Byline Burpy," *Arrow Lakes News*, 31 March 1971, 2; "Campbell Asks Hydro to Remove Stumps," *Arrow Lakes News*, 1 November 1971, 1.

59 "Byline Burpy," *Arrow Lakes News*, 4 April 1972, 4, reflecting on spring and summer of 1972; "High and Dry," *Arrow Lakes News*, 11 July 1973, 2; "Arrow Reservoir Now at Peak for the Summer," *Arrow Lakes News*, 18 July 1973; "Clean Up the Valley," *Arrow Lakes News*, 25 July 1973, 11; "Nakusp Beach Closes Due to Receding Lake," *Arrow Lakes News*, 25 July 1973, 1; "Arrow Lakes Desert," *Arrow Lakes News*, 25 July 1973, 14; "Still No Safe Beach Found in Nakusp," *Arrow Lakes News*, 1 August 1973, 1; "Byline Burpy," *Arrow Lakes News*, 1 August 1973, 2; "Rec Commission Raps BC Hydro," *Arrow Lakes News*, 3 October 1973, 1.

60 Bob Harrington, "It's Your World," *Arrow Lakes News*, 29 August 1973, 3; Editorial, "Keep Level Below Peak," *Arrow Lakes News*, 17 July 1974, 2; "Byline Burpy," *Arrow Lakes News*, 24 July 1974, 2; photo and caption, *Arrow Lakes News*, 1 May 1974, 1.

61 Interview of William Barrow; "Byline Burpy," *Arrow Lakes News*, 24 November 1971, 2; Ibid., 17 October 1973.

62 Interview of Ernie Orr; "Byline Burpy," *Arrow Lakes News*, 23 July 1975, *Arrow Lakes News*, photo and caption, 10 July 1974, 7; "Byline Burpy," *Arrow Lakes News*, 28 March 1973, 2; "Reservoir Stumps Overturn Boat," *Arrow Lakes News*, 26 July 1972, 10.

63 *Arrow Lakes News*, 16 April 1975, 1.

64 "Raise up Our Reservoir," and "Byline Burpy," *Arrow Lakes News*, 21 May 1975, 2; "Hydro Sidesteps Lake Level issue," *Arrow Lakes News*, 18 June 1975, 1.

65 "Raise up Our Reservoir."

66 Julia Kristeva, "Approaching Abjection," in *Powers of Horror: An Essay on Abjection*, trans. Leon S. Roudiez (New York: Columbia University Press, 1982), chap. 1.

67 Goldsmith and Hildyard, *Social and Environmental Effects*, 17; Wilson and Conn, "On Uprooting," 44.

68 René Descartes, *The Philosophical Writings of René Descartes*, vol. 2, *Meditations on First Philosophy*, trans. J. Cottingham, R. Stoothoff, and D. Murdoch (Cambridge: Cambridge University Press, 1984 [1641]), 2:1-62; Casey, *Remembering* and *Getting Back into Place*.

69 "Local Artist Jean Spicer, Nakusp, BC," *Arrow Lakes News*, 5 July 1972, 8; "'Doggy Socks' Prove Initiative of BC Women," *Arrow Lakes News*, 7 August 1941, reprinted from the *Vancouver Province*, 20 July 1941; "Golden Wedding," *Arrow Lakes News*, 20 July 1967, 1. See Stuart McLean, *Welcome Home: Travels in Small Town Canada* (Toronto: Viking, 1992), 349.

70 Arrow Lakes Historical Society, Living History Project Interview, 1974.

71 Waterfield, *Continental Waterboy*, 62.

72 McLean, *Welcome Home*, 348.

73 Ibid.

74 "Miss Waterfield Won Highest Aggregate in Flower Show," *Arrow Lakes News*, 23 August 1956, 1; "Women's Institute Flower Show at Nakusp Very Respectable Show," *Arrow Lakes News*, 15 August 1957, 3; "Quality High in Flower Show, Nakusp, Mrs Spicer Top Winner," *Arrow Lakes News*, 18 August 1960, 4. There are pictures of Jean's gardens before the

inundation and of the rose garden she and Chris made along the front thereafter at http://megaprojects.uwo.ca/ArrowLakes.

75 *Arrow Lakes News,* 20 September 1956, 19 September 1957; Waterfield, *Continental Waterboy,* 64.

76 *Arrow Lakes News,* 31 August 1988; 5 May 1991.

77 Chris Spicer, "Letter to the Editor," *Arrow Lakes News,* 17 April 1974, 2.

78 "Harvesting Nakusp Soil for 20 years," *Arrow Lakes News,* 31 August 1988, 5.

79 Jean Spicer's strongest public statements on animal welfare were made after the flooding and were concerned with the effects of the flooding on wildlife. See "Letters to the Editor," *Arrow Lakes News,* 23 October 1968, 6; 20 September 1972, 3; 4 July 1973, 2; photo, 31 July 1974.

80 *Arrow Lakes News,* 31 August 1988, 5; John Porter, *The Vertical Mosaic: An Analysis of Social Class and Power in Canada* (Toronto: University of Toronto Press, 1965); John Seeley, *Crestwood Heights* (Toronto: University of Toronto Press, 1956); Geraldine Pratt, "Home Decoration and Expression of Identity" (MA thesis, University of British Columbia, 1980).

81 *Arrow Lakes News,* 31 August 1988, 5.

82 "Hydro Plans for Waterfront Given," *Arrow Lakes News,* 10 March 1966, 1; Waterfield, *Land Grab,* 95-97; Wilson, *People in the Way,* 157-59.

83 "Why No Local Produce?" *Arrow Lakes News,* 23 August 1972.

84 *Arrow Lakes News,* photo, 29 March 1973; photo, 25 April 1973, 1; photo, 11 July 1973, 8; "Byline Burpy," *Arrow Lakes News,* 17 July 1974, 2; *Arrow Lakes News,* photo, 1 October 1975, 1; *Valley Voice,* 20 August 1998.

85 Arundhati Roy, *The Cost of Living* (New York: Viking, 1999); Mark Fiege, *Irrigated Eden: The Making of an Agricultural Landscape in the American West* (Seattle: University of Washington Press, 1999); Keith Petersen, *River of Life, Channel of Death: Fish and Dams on the Lower Snake* (Corvallis: Oregon State University Press, 1995); Harden, *A River Lost;* McCully, *Silenced Rivers;* Eric Swyngedouw, "Modernity and Hybridity: Nature, Regeneracionismo, and the Production of the Spanish Waterscape, 1890-1930," *Annals of the Association of American Geographers* 89, 3 (1999): 91-102; Sean McCutcheon, *Electric Rivers: The Story of the James Bay Project* (Montreal: Black Rose Books, 1991); Richard Salisbury, *A Homeland for the Cree: Regional Development in James Bay, 1971-1981* (Montreal: McGill-Queen's University Press, 1986); James Hornig, *Social and Environmental Impact of the James Bay Hydroelectric Project* (Montreal: McGill-Queen's University Press, 1999).

86 *The Columbia Kootenay Symposium,* videorecording, (Nelson, BC: Shaw Cable 10, 1993).

87 "Spicer Centre," *Arrow Lakes News,* 8 September 1999; Western Sustainable Agriculture Working Group, http://www.westernsawg.org; "To the Select Standing Committee on Agriculture and Fisheries," 26 November 1999, http://www.fooddemocracy.org/docs/SAWG-BC_Nov99.pdf; Hollyhock, http://www.hollyhock.ca/cms; 1998/99 Legislative Session: 3rd Session, 36th Parliament: Select Standing Committee on Agriculture and Fisheries, November 4, 1999, http://www.leg.bc.ca/CMT/36thParl/CMT08/hansard/1999/af110499.htm.

CHAPTER 6: SMELL AND RISK

1 Using these same audio excerpts about knowledge of danger, I explore the emotional content of their pauses and inflection belying the words used as what is "lost in transcription"

in "Lost in Transcription: Oral and Textural Readings of Interviews from Bruce County" (ms).

2 Mary Douglas, *Purity and Danger: An Analysis of Concepts of Pollution and Taboo* (New York: Praeger, 1966).

3 Fabriziomaria Gobba, "Occupational Exposures to Chemicals and Sensory Organs: A Neglected Field of Study," *Neurotoxicology* 24 (August 2003): 678.

4 Alain Corbin, *The Foul and the Fragrant: Odor and the French Social Imagination* (Cambridge: Harvard University Press, 1986), 6; Paul Rodaway, *Sensuous Geographies: Body, Sense and Place* (New York: Routledge, 1994), 70; Constance Classen, David Howes, and Anthony Synnott, *Aroma: The Cultural History of Smell* (New York: Routledge, 1994), 3, 5; H.T. Lawless and T. Engen, "Associations to Odors: Inference, Mnemonics and Verbal Labelling," *Journal of Experimental Psychology* 3, 1 (1977): 52-59; Lisa M. Mitchell and Alberto Cambrosio, "The Invisible Topography of Power: Electromagnetic Fields, Bodies and the Environment," *Social Studies of Science* 27, 2 (1997): 226; John Urry, "Sensing the City," in Dennis R. Judd and Susan Fainstein, eds., *The Tourist City* (New Haven: Yale University Press, 1999), 81.

5 Alain Corbin, *Time, Desire and Horror: Towards a History of the Senses* (Cambridge, MA: Polity Press, 1995), 191; Fernando Coronil, "Smelling Like a Market," *American Historical Review* 106 (February 2001): 119-29; Joy Parr, "Local Water Diversely Known: Walkerton 2000 and After," *Environment and Planning D: Society and Space* 23, 2 (2005): 2; Connie Y. Chiang, "Monterey-by-the-Smell: Odors and Social Conflict on the California Coastline," *Pacific Historical Review* 73 (May 2004): 184; Mark S.R. Jenner, "Civilisation and Deodorisation? Smell in Early Modern English Culture," in Peter Burke, Brian Harrison and Paul Slack, eds., *Civil Histories: Essays Presented to Sir Keith Thomas*, (Oxford: Oxford University Press, 2000), 138.

6 Corbin, *Foul and Fragrant*, 7; J. Douglas Porteus, "Smellscape," *Progress in Geography* 9 (September 1985): 358-59; Mark M. Smith, "Making Sense of Social History," *Journal of Social History* 37 (Fall 2003): 165-86; Classen, Howes, and Synnott, *Aroma*, 57, 170-71; Jenner, "Civilisation and Deodorisation," 133.

7 Corbin, *Time, Desire and Horror*, 182; Classen, Howes, and Synnott, *Aroma* 8; Chiang, "Monterey," 185.

8 Gerald Killan, *Protected Places: A History of Ontario's Provincial Parks System* (Toronto: Dundurn Press, 1995), 102, 124.

9 For a history of prosecutions under the act, see Dianne Saxe, "Fines Go Up Dramatically in Environmental Cases," *Canadian Environmental Law Reports N.S* 3 (1989): 104.

10 Canadian Centre for Occupational Health and Safety, *Hydrogen Sulphide* (Hamilton, ON: The Centre, 1985), 13. This document is in the *Chemical Hazard Summary Series* in microform (Toronto: Micromedia, 1994), microlog: 87-00815.

11 Canadian Centre for Occupational Health and Safety, *Hydrogen Sulphide*, 4; Rhoderic J. Reiffenstein, W.C. Hulbert, and S.H. Roth, "Effects of Hydrogen Sulphide (H_2S) on Humans and Animals," *Annual Reviews of Pharmacology and Toxicology* 32 (1992): 111.

12 Wladimir Paskievici and L. Zikovsky, "Public Health Risks Associated with the CANDU Nuclear Fuel Cycle: Non-Radiological Risks" (Ottawa: Atomic Energy Control Board, September 1982), 37-39, available on microfiche (Toronto: Micromedia, 1994), microlog 85-02818. Canadian Centre for Occupational Health and Safety, *Hydrogen Sulphide*, 4-6, 13; Reiffenstein, Hulbert, and Roth, "Effects of Hydrogen Sulphide," 112-13. On somatic

effects in the Alberta sour gas fields, see Andrew Nikiforuk, *Saboteurs: Wiebo Ludwig's War against Big Oil* (Toronto: Macfarlane Walter and Ross, 2002).

13 On a similar scenario of eroding trust, see Sheila Jasanoff, "Civilisation and Madness: The Great BSE Scare of 1996," *Public Understanding of Science* 6, 3 (1997): 221-32.

14 Robert Bothwell, *Nucleus: The History of Atomic Energy Canada Limited* (Toronto: University of Toronto Press, 1988), 197-211; Robin Cowan, "Nuclear Power: A Study in Technological Lock-in," *Journal of Economic History* 50, 3 (1990): 545-47. Heavy water, deuterium oxide, occurs naturally at a ratio of one part to seven thousand in ordinary water (H_2O). It contains two deuterium rather than two hydrogen atoms and is twice as heavy. In a nuclear reaction, heavy water attracts fewer neutrons than light water and as a moderator will thus sustain a chain reaction with natural uranium, an element that during fission releases fewer neutrons than enriched uranium. Whereas builders of light-water reactors used abundant ordinary water and needed supplies of expensive enriched uranium, builders of heavy water reactors could use natural uranium, which was abundant in Canada, but needed sources of expensive heavy water.

Heavy water for the first seven Canadian reactors, including the one at Douglas Point, came from Savannah River, Georgia, through an agreement with the American Atomic Energy Commission. In the 1960s, with the US capacity to produce heavy water reduced and an Ontario Hydro decision to aggressively expand nuclear generating capacity in place, Atomic Energy Canada Limited (AECL), the federal agency in charge of the CANDU program, contracted with two private firms to build heavy water plants. Both were slow to come on line. Thus, with some urgency, in 1969 AECL sought approval to build a heavy water plant at the Douglas Point nuclear site, where the existing small reactor presently was to be joined by two of more commercial scale. Through the 1970s, as the OPEC crisis drove up costs at fossil fuelled power plants, the acute shortage of heavy water constrained the start-up schedules for new nuclear generating stations. Consistent with Canadian federal government economic development priorities, both were located in struggling coal-mining and steel-making communities in Nova Scotia. One of these, at Glace Bay, designed to use sea water, utterly failed. The other, built by Lummis Canada, a subsidiary of the designer of Savannah River, was producing at half capacity until 1974. In 1972, with Canadian heavy water production facilities still prospective, heavy water had to be begged, borrowed, and bought from Argentina, Sweden, the United Kingdom, Russia, and the United States. Subsequently, the heavy water to commission new units was transferred from the research and demonstration reactors, which were shut down on rotation until the heavy water supply crisis could be resolved. See H.K. Rae, *Canada Enters the Nuclear Age: A Technical History of Atomic Energy of Canada Limited* (Kingston and Montreal: McGill-Queen's University Press, 1997), 334, 337, 339. The local press reported that the initial loading alone of the eight reactors under construction and planned for the Bruce site through the 1970s and early 1980s would require 6,400 tons of heavy water. See *Kincardine News*, 19 January 1972, 8 March 1972, 29 March 1972, 25 April 1973, 19 July 1974, and 18 September 1974; *London Free Press*, 2 March 1972, 6 March 1972, 29 March 1972, 11 August 1973, and 18 January 1974.

15 Here the insights of Ulrich Beck and Frank Fisher apply. See Ulrich Beck, *Ecological Enlightenment: Essays on the Politics of the Risk Society* (Atlantic Highlands, NJ: Humanities International Press, 1995), 104-5; and Frank Fisher, *Citizens, Experts and the Environment: The Politics of Local Knowledge* (Durham, NC: Duke University Press, 2000), 54.

16 Paskievici and Zikovsky, "Public Health Risks," 70.

17 Rae, "Heavy Water," in *Canada Enters the Nuclear Age,* 334-35; "Bruce Heavy Water Plant, a Short Description of the Process Involved, 23 June 1969," Library and Archives Canada (LAC), Atomic Energy Control Board, RG 29 12 1106-14-1.2. A G-S process tower contains many sieve trays, with a cold section in the top half and a hot section below. Feed water at the Bruce site, water from Lake Huron, entered from the top and passed over a series of plates. At each pass through a tray in the cold section, deuterium was exchanged from the gas to the water so that the water picked up more deuterium as it moved down the tower. At the bottom of the cold section, some enriched water was withdrawn to be further enriched in another tower. The remainder passed into the hot section, where deuterium was stripped from the water, providing enriched gas to feed the cold section. The depleted water was drawn from the bottom of the tower, passed through H_2S strippers, and returned to the lake. The enriched gas passed from the top of the tower on to a second stage to be processed again. Excess H_2S was vented directly into the atmosphere from the top of the cold tower.

18 Scales formed blocking flow through the trays; gas moving at higher velocities in the towers corroded the carbon steel portions of plant pipes, particularly at the bends, causing unplanned ground level releases of H_2S through pin-hole leaks. See Rae, "Heavy Water," 339-45; interview of Jim Dalton, Kincardine, ON, 3 November 2001. Dalton was Lummis project head for the Bruce Heavy Water Plant.

19 AECB staff annual assessment of Bruce Heavy Water Plant, 1997 (Ottawa: AECB, 1998), iii, in microform (Toronto: Micromedia 1999), microlog 99-07311; Michael Prior, Michelle Mostrom, Robert Coppock, and Zack Florence, *Environmental Health Scoping Study at Bruce Heavy Water Plant,* AECB Project 3.168.1 (Ottawa: AECB, October 1995), 35-36. For H_2S emissions for the 1990s, see AECB staff annual assessment of the Bruce Heavy Water Plant, 1996 (Ottawa: AECB, 1997), 4, in microform (Toronto: Micromedia, 1998), microlog 98-00944.

20 J. Douglas Porteous, "Smellscape," *Progress in Geography* 9, 3 (1985): 358; Rodaway, *Sensuous Geographies,* 68.

21 Mean wind speed was 20 km/hour or greater from October through January; Michael Prior et al., *Environmental Health Scoping,* 36, 40; Paskievici and Zikovsky, "Public Health Risks," 73.

22 Interview of Jim Dalton.

23 Paskievici and Zikovsky, "Public Health Risks," 73, 77; LAC, RG 29 7 1106-14-3 (2), Bruce Heavy Water Plant Safety Advisory Committee (BHWPSAC), "An Evaluation of the Impact of the Plan on Health and Safety of the Surrounding Area," October 1970, 7.

24 In 1982, the number one recommendation by scientists charged to examine non-radiological public health risks associated with the full CANDU nuclear fuel cycle was the "establishment of good theoretical plume rise formulas for (heavy water plant gaseous emission in) unstable weather classes." See Paskievici and Zikovsky, "Public Health Risks," 99.

25 The board was wary of pressure from the AECL and Ontario Hydro to speedily license the Bruce heavy water plant. Board members knew from the experience with leaks, accidents, and "knock-downs" that heavy water plants at commercial scale were hazardous to people "within the range of influence" of the plant. See LAC, RG 29 12 1106-14-3, Minutes of the first meeting of the Heavy Water Plant Safety Advisory Committee (HWPSAC) 19 March 1969, 1, 2, 7.

26 Perhaps not surprisingly, given their formation in the universities and public services of the Commonwealth, "the cultural characteristics of caution, empiricism and restraint" that

Jasanoff found "imprinted on the provision of scientific advice bearing on health, safety and environmental risks" in Britain are apparent in the work of the board. Jasanoff, "Civilisation and Madness," 228.

27 Between the AECB and Ontario Hydro, the thorniest issue, never resolved to the satisfaction of some experts on the advisory committee, was the relationship between safety analysis and the design of the plant. The board would have preferred the Bruce facility to be re-engineered on the basis of the daunting earlier Canadian experience. The AECL emphasized that adopting an existing design and proceeding rapidly with construction was the only way to meet the "urgent need for heavy water," which meant, in effect, as Ulrich Beck has observed, that "the open air of daily life" would constitute the laboratory in which "the properties, safety and long-term effects" of the technology would be studied. See LAC, RG 29 12 1106-14-3, F.C. Boyd to J.F. Foster, 8 May 1969, and J.F. Foster to F.C. Boyd, 8 August 1969; Fisher, *Citizens, Experts,* 54; Beck, *Ecological Enlightenment,* 104-5.

28 LAC, RG 29 12 1106-14-3, 1 and 12, attachments to J.S. Foster, AECL to F.C. Boyd, HWP-SAC, 8 July 1969, note 1 and note 3, 9 June 1969; minutes of HWPSAC private meeting, 30 April 1969; LAC, RG 29 7 1106-14-3 (2), report of the BHWPSAC, "An Evaluation of the Impact," 8 October 1970, 8; BHWPSAC, Final Report, October 1971, 4, 5.

29 LAC, RG 29 12 1106-14-3 1, HWSAC minutes of 1 October 1969; Michele Murphy, "The 'Elsewhere within Here,' and Environmental Illness; or, How to Build Yourself a Body in a Safe Space," *Configurations* 8, 1 (2000): 101; Mitchell and Cambrosio, "Invisible Topography," 228, 239; and Allan Mazur, *Hazardous Inquiry: The Rashomon Effect at Love Canal* (Cambridge: Harvard University Press, 1998), 58, 205, 230.

30 LAC, RG 29 12 1106-14-3 1, A.K. DasGupta to R.M. Duncan, Secretary of the HWPSAC, 24 July 1969.

31 LAC, RG 29 7 1106-14-3 (2), report of the BHWPSAC, "An Evaluation of the Impact," 7, 11, and unpaginated summary. DasGupta was one of three signatories to this report. See LAC, RG 29 7 1106-14-3 2, A.K. DasGupta to Dr. A.H. Booth, Assistant Director, Environmental Health, Department of Health and Welfare, 19 April 1971.

32 One commissioned research group modelled detonable and flammable gas clouds in the event of catastrophic failures in H_2S storage and then refined this model for forty-three accidental release scenarios at the Bruce plant. Another, which included members of the Defence Research Establishment at Suffield, Alberta, conducted experimental tests on the detonation of H_2S in the open air and of its flame acceleration and detonation in confined environments. See A.J. Saber, *Investigation of the Explosion Hazards of Hydrogen Sulphide, Phase I* (Ottawa: AECB, 1986), in microform (Toronto: Micromedia, 1994), microlog 86-05414 1, 3, 4, 7; I.O. Moen, *Investigation of the Explosion Hazards of Hydrogen Sulphide, Phase II* (Ottawa: AECB, 1986), in microform (Toronto: Micromedia, 1994), microlog 88-06131 1, 2, 6.

33 On dilemmas, see Allan Mazur, *True Warnings and False Alarms: Evaluating Fears About the Health Risks of Technology, 1948-1971* (Washington, DC: Resources for the Future, 2004), 4-5, 52; Joel Tarr, *The Search for the Ultimate Sink: Urban Pollution in Historical Perspective* (Akron, OH: University of Akron Press, 1996), 14; Fisher, *Citizens, Experts,* 52; M. Thompson, "The Management of Hazardous Wastes and the Hazards of Wasteful Managment," in *Dirty Words: Writings on the History and Culture of Pollution,* ed. Hannah Bradby, (London: Earthscan 1991), 118-21; Stephen Bocking, *Nature's Experts: Science, Politics and the Environment* (New Brunswick, NJ: Rutgers University Press, 2004), 136-37.

34 Archives of Ontario (hereafter AO), Department of Energy and Resources Management, RG 12 7-1 cont 9, file "Bruce Heavy Water Plant, Douglas Point," W.A. Neff to W.B. Drowling, Air Management Branch, 27 July 1971.

35 Gobba, "Occupational Exposures," citing A.R. Hirsch and G. Zavala, "Long-Term Effects on the Olfactory System of Exposure to Hydrogen Sulphide," *Occupational and Environmental Medicine* 56 (1999): 284-87; Ulrich Beck, *Ecological Politics in an Age of Risk* (London: Polity, 1995), 56-57; Interview of Barry Schell, Southampton, 15 October 2001.

36 Interview of Ben Cleary, Kincardine, ON, 27 October 2001. The improvements he noted were better seals and improved compressor control, superior welding techniques and stress relieving, greater attention to inspection, and technical assurance programs.

37 Paskievici and Zikovsky, "Public Health Risks," 77, 99.

38 "Just Like the Old Family Fridge," *Kincardine News*, 12 January 1972; "The Lousy Smell, Is It Dangerous? What Is It? When Will It End?" *Kincardine News*, 20 December 1972.

39 Chiang, "Monterey," 213; Tarr, *Search for the Ultimate Sink*, 15.

40 Interview of John MacKenzie, Underwood, ON, 3 October 2001; Eric Howald, "Editorial: A Blessing or Bane," *Kincardine News*, 14 March 1973.

41 Interview of Eldon Roppel, Tiverton, ON, 15 October 2001; interview of James Weir, Tiverton, ON, 4 October 2001.

42 Interviews of Jim Dalton and Eldon Roppel.

43 Samuel P. Hayes, *Beauty, Health and Permanence: Environmental Politics in the United States* (New York: Cambridge University Press, 1987), 172-73.

44 Interview of Jim Dalton.

45 Interview of Ben Cleary.

46 Interview of Jim Dalton.

47 William Leiss, "Between Expertise and Bureaucracy: Risk Management Trapped at the Science-Policy Interface," in G. Bruce Doern and Ted Reed, eds., *Risky Business: Canada's Changing Science-Based Policy and Regulatory Regime* (Toronto: University of Toronto Press, 2000), 49-50; M. Thompson, "Management of Hazardous Wastes," in *Dirty Words*, 121; Bocking, *Nature's Experts*, 138.

48 AO, RG 1-47-3, cont 45, file 1973 27-0702-1, "Lease, Inverhuron Provincial Park Ontario Hydro," 11 March 1974. The lease dated 24 July 1973 was given provincial cabinet approval on 8 November 1972. See AO, RG 1-121 A151.10, acc. 150 52/9, box 7, Inverhuron Park – operations – 1975 no. 255, "Event Sequence, Inverhuron Park Problem."

49 The lease was not made public, and many of the cottagers first saw it when I removed it from my file at the end of our interviews in 2001. Robert Mackenzie referred to the inaccessibility of the lease during our interview.

50 Interviews of Ben Cleary and Robert Wilson, Scarborough, ON, 1 November 2001.

51 Jasanoff, "Civilization and Madness," 228.

52 *London Free Press*, 3 September 1973, 1; 3 October 1973, 1; 28 July 1975, 8.

53 *Kincardine News*, 3 May 1972, 1; "Bruce Plant to Test Poison Gas Warning," *London Free Press*, 30 September 1972; AO, RG 1-8 acc. 15052, vol. 4, HEPC-BWHP-MacGregor Point, George Gathermole, Chairman, Hydro-Electric Power Commission of Ontario, to Leo Bernier, Minister of Natural Resources, 25 July 1972; "Poison Gas Shelters Planned at Inverhuron Park," *London Free Press*, 15 June 1972; "Five Shelters at Inverhuron Park," *Kincardine News*, 11 April 1973; AO, RG 1-47-3, cont 65, file 27-0702, "1975 Budget of Costs to be Assumed by Ontario Hydro," and "Inverhuron Provincial Park Future Management Proposal," 3 October 1975.

54 On the dilemmas of applying the precautionary principle to uncertain hazards, see Bocking, *Nature's Experts,* 148. The opening of the replacement camping park to the north at MacGregor Point was delayed until 1976 by prolonged negotiations with the United Auto Workers, a union that, for decades, had used the MacGregor Point site as a recreational and educational centre for its members. See AO, RG 1-47-3, cont 65 27-0702, R.H. Hambly, Park Management Branch to W. Charlton, Regional Director, 14 March 1975, Inverhuron Park (Confidential); AO, RG 1-121 A151.10, box 7, Inverhuron Park-operations-1975 no. 255, D.K. Reynolds, Deputy Minister, Natural Resources, to Leo Bernier, Minister, 4 February 1975, "Expropriation of United Auto Workers Property."

55 AO, RG 1-8, acc. 15052, vol. 4, HEPC-BHWP-MacGregor Point, Fritz Knechtel, Hanover, to Leo Bernier, 7 March 1973; *London Free Press,* 24 September 1973, 8; *Kincardine News,* 20 March 1974, report of a public meeting for residents of Bruce Township, which surrounded the nuclear and heavy water plants; "Details of Development Freeze Still Unknown," *Kincardine News,* 14 November 1973; "Bruce Township Restrictions Explained," *Kincardine News,* 28 November 1973.

56 AO, RG 1-8, acc. 15052, vol. 4, HEPC-BWHP-MacGregor Park, James Auld to G.E. Gathermole, copied to Darcy McKeough and Leo Bernier, 28 September 1973. The files show officials struggling with the conflicting interests of a "natural" resource, here the parksite, for recreation and for power generation, a dilemma akin to that Jasanoff observed in the British BSE crisis of 1996 as a single Ministry of Agriculture, Fisheries, and Food was charged to balance public health and the economic viability of the agricultural sector. See Jasanoff, "Civilisation and Madness," 225-26.

57 AO, RG 72-55 1 of 2, Ontario Hydro 1973 Bruce Generating Station, D.P. Caplice to K.H. Sharpe, 22 October 1973; AO, RG 1-8, acc. 15052, vol. 4, HEPC-BHWP-MacGregor Point 1973, R. Hambly to P. Addison, 5 May 1971; AO, RG 12-45 1972 811-6, vol. 279, BWHP 1972, "Statement on the Bruce Heavy Water Plant," and C.J. Macfarlane, Air Management Branch to W.B. Drowley, Air and Pollution Control Division, 29 June 1972.

58 "Inverhuron Safety Plan Explained," *Kincardine News,* 30 August 1972.

59 "Park Not Closing ... Yet," *Kincardine News,* 17 January 1973, 1; "Inverhuron: Woods, Beach – and Poison Gas Next Door," *London Free Press,* 30 June 1972, 39, and 9 July 1973, 5.

60 There is a description of a similar contradiction in Mitchell and Cambrosio, "Invisible Topography." The quoted phrase is from Fisher, *Citizens, Experts,* 51.

61 "Danger Limits around Heavy Water Plant Explained," *Kincardine News,* 17 October 1973; "Kincardine Told Gas Leakage Safe," *London Free Press,* 24 September 1973, 8, and sec. 2, 25.

62 "Group Says Hydro Polluting Bruce Area," *London Free Press,* 26 September 1973; "Progress Report on Foul Odor Hunt Refused by Hydro," *London Free Press,* 2 October 1973, 25; "Battle to Save Park Lost," *Kincardine News,* 17 October 1973.

63 "The Story behind Inverhuron Park," *Kincardine News,* 7 January 1976; on the archaeological sites in Inverhuron Park, see J.V. Wright, *Knechtel 1 Site, Bruce County, Ontario* (Ottawa: National Museum of Man, 1972).

64 Fritz Knechtel, "Save Inverhuron Park," *London Free Press,* 11 July 1973; Fritz Knechtel, "Our Readers Say," *Kincardine News,* 19 September 1973; AO, RG 1-8, acc. 15052, vol. 4, HEPC-BHWP-MacGregor Point, Fritz Knechtel to James Auld, Minister of the Environment, 29 September 1973, and Auld's reply, 2 November 1973. There are similar recollections from other park users: interviews of Gerald Paul, Kingston, 20 November 2001 (Paul was the minister at the United Church in Kincardine from 1977, a bird-watcher, naturalist,

and essayist); Robert Wilson (Wilson, the Glasgow-trained physicist and radiation protection manager at Bruce A, later a senior occupational health and safety officer with wide responsibilities at Ontario Hydro, was, during his time at Bruce, an active member of the Kincardine Library Board); William MacKenzie, Southampton, ON, 22 October 2001 (MacKenzie had grown up in Tiverton and worked as an attendant at Inverhuron while earning a geography degree at the University of Western Ontario. He was later an alderman in Kincardine and a nuclear operator at Bruce.)

65 Bocking, *Nature's Experts*, 144, 148-53; Jasanoff, "Civilisation and Madness," 229, 231.

66 AO, RG 1-8, acc. 15052, vol. 4, HEPC-BHWP-MacGregor Point, Leo Bernier to George Gathermole, 31 May 1973, and "Inverhuron Park—Chronological Summary," item 12; "Compliant Response Report, Bruce Heavy Water Plant Douglas Point in Response to William and June Ruddock."

67 After being forwarded both a telephone complaint from William Ruddock to the Air Management Branch and his wife June Ruddock's letter, Leo Bernier, minister of natural resources, pressed George Gathermole, chairman of Ontario Hydro, to make a double-barrelled public announcement on 7 June 1973. The four reactors being built at the Bruce, known as Bruce A, would presently be joined by a companion, the four new reactors that became Bruce B, and a second installation at the heavy water plant capable of producing eight hundred tons of heavy water a year. Construction of both these new facilities began in 1976, at an estimated cost of $1.5 billion, with the first reactor in the new station scheduled for completion by 1978. In order "to provide Hydro with the necessary green belt to surround the enlarged nuclear hydro facilities," all camping would be moved to MacGregor Point, with the portion of Inverhuron outside the green belt to surround the enlarged nuclear hydro facilities accessible for day-use swimming and picnicking. Although more northerly portions of the Bruce nuclear site were unoccupied, like the heavy water plants, the Bruce B generating station would be built on the southern boundary of the property, adjacent to Inverhuron Park. See AO RG 1-8, acc. 15052, vol. 4, HEPC-BHWP-MacGregor Point, Leo Bernier to George Gathermole, 31 May 1983.

68 June Ruddock was a long-time member of the Ontario Federation of Naturalists and involved with the Algonquin Wilderness League. Interview of Frank Ruddock, Ottawa, 18 November 2001. The letters of both writers were rich with allusions to other parks that had been and might be threatened by industry. See AO RG 1-8, acc. 15052, vol. 4, HEPC-BHWP-MacGregor Point, June Ruddock to Leo Bernier, 22 June 1973; Knechtel letters, *London Free Press*, 11 July 1973, and *Kincardine News*, 19 September 1973, See also a similar editorial, "Can the Park Be Saved," *Kincardine News*, 13 June 1973.

69 Interview of Robert Mackenzie, Tiverton, 17 October 2001.

70 Fisher, *Citizens, Experts*, ch. 7; Jasanoff, "Civilisation and Madness," 223; Brian Wynne, "May the Sheep Safely Graze," in *Risk, Environment and Modernity: Towards a New Ecology*, ed. Scott Lash, Bronislaw and Brian Wynne (London: Sage, 2004), 45, 61.

71 Interviews of Robert MacKenzie, Inverhuron, ON, 17 October 2001; interview of Frank Ruddock. AO RG 1-8, acc. 15052, vol. 4, HEPC-BHWP-MacGregor Point, Christine Feaver, Hamilton, ON, to Leo Bernier, 18 June 1973.

72 The exception was an oil executive, Ray Schumacher. Interview of Robert Mackenzie.

73 William Ruddock, "Rejects Inverhuron Editorial," *London Free Press*, 12 July 1973, 6; William Ruddock, "Letter to the Editor," *Kincardine News*, 5 September 1973; William Ruddock, "Inverhuron Park," *London Free Press*, 31 August 1973, 7; AO RG 1-8, acc. 15052, vol.

4, Inverhuron Park 1973, Inverhuron Committee of Concern, William C. MacKenzie to William G. Davis, Premier, 10 July 1973; "Our Readers," *Kincardine News*, 11 July 1973; "NDP Leader Speaks on Inverhuron," *Kincardine News*, 21 November 1973, 11.

74 AO, RG 12-45 544-4, vol. 296, Mike Singleton, Federation of Ontario Naturalists, to James Auld, 24 July 1973; C.J. Macfarlane, Director, Air Management Branch, to Everett Bigg, Deputy Minister, Ministry of the Environment, 7 August 1973; James Auld to Mike Singleton, 10 August 1973; Mike Singleton to James Auld, 21 August 1973; C.J. Macfarlane to Mike Singleton, 13 September 1973. Macfarlane assured the Ontario naturalists that a continuous air quality monitoring system for SO_2 and H_2S would shortly be established in Inverhuron Village and the data made available to the public on request.

75 Interview of Frank Ruddock.

76 "Inverhuron Protestors Seek Probe on Takeover," *London Free Press*, 14 July 1973, 1; "Save Inverhuron Committee Formed," *Kincardine News*, 4 July 1973, 1; "Camp Delights, Hydro Needs Conflict at Inverhuron Park," *Globe and Mail*, 25 June 1973, 5. There are copies of the pamphlets in AO, RG 1-8, acc. 15052, vol. 4, file Inverhuron Park 1973. See interview of Frank Ruddock; AO, RG 12-45 811-6 1973, cont 305, Jeanne Quayle, London to Ministry of Environment, 18 June 1973. See also "Our Readers Say," *Kincardine News*: Bill Mackenzie, 13 June 1973, 4; Barbara Macklem, 4 August 1973; Fritz Knechtel letter, 15 August 1973. See "Inverhuron Protest Growing, 4,000 Sign Petition against Hydro Takeover," *London Free Press*, 8 August 1973, sec. 3, 1; "Our Readers Say," *Kincardine News*, 8 August 1973 and 15 August 1973; "Editor's Notebook," 27 June 1973; "Inverhuron Committee Charges Hydro with Misrepresentation," 25 July 1973, 1; "Committee: Hydro Wants Park for Reactor, not Gas Danger," 15 August 1973, 1; "No Part Acquisition, No Plant, McKeough Warns," *London Free Press*, 21 August 1973, 3, 4; "Battle Lost over Inverhuron Camping," *Globe and Mail*, 11 October 1973.

77 AO, RG 1-47-3, vol. 65 27-0702, M.C. White, Senior Project Officer, Heavy Water Plants, to J.W. Keenan, Executive Director, Division of Parks, Ministry of Natural Resources, 17 March 1976.

78 Mazur, *Hazardous Inquiry*, 39, 48, 58, 205, 230; Murphy, "The 'Elsewhere within Here,'" 88.

79 Christopher Sellers, "Thoreau's Body: Towards an Embodied Environmental History," *Environmental History* 4, 4 (1999): 486.

80 Murphy, "The 'Elsewhere within Here,'" 114-15.

81 Interview of Eugene Bourgoise, Tiverton, 2 October 2001.

82 "The Effects of Gaseous Emissions from the Bruce Heavy Water Plant on a Local Sheep Farmer" (Ottawa: Atomic Energy Control Board, 3 October 1995), BMD 95-143; M. Slana, "Report on Assessment of Lamb Mortality Data from Bourgeois Farm," Ontario Veterinary College, University of Guelph, April 1993, as Exhibit 10, fax header from Atlantic Veterinary College, Descriptive Statistics section, on Eugene Bourgeois Fumigaton. CD available from Eugene Bourgeois, Tiverton, ON.

83 R.J. Raffenstein, Department of Pharmacology, University of Alberta, 21 November 1991, Exhibit 12, Eugene Bourgeois document CD; Prior et al., *Environmental Health Scoping*, 20-23.

84 "The Effects of Gaseous Emissions," BMD 95-143. The "minor but measurable" quote is on p. 5. See also Ontario Ministry of Environment and Energy, Technical Memorandum, Report No: SDB-008-3511-95TH, "Phytotoxicology Vegetation Study: E Bourgeois—Tiverton (1994)," R. Emerson to R. Pearson, 16 March 1995, on Bourgeois fumigation CD;

interview of Eugene Bourgeois; "Heavy Water Plant Safety: The Siting Guidelines Action taken in regard to Mr. E. Bourgeois' Problems" (Ottawa: Atomic Energy Control Board, 1 August 1990), BMD 90-139.

CHAPTER 7: TASTE AND EXPERTISE

1 http://www.wefeelfine.org.
2 Pierre Payment, 28 February 2001; James Kiefer, 16 November, 2000. http://mail.tscript.com/trans/walkerton2000.htm.
3 The events of these days are clearly chronicled in Colin Perkel, *Well of Lies: The Walkerton Tragedy* (Toronto: McClelland and Stewart, 2002), 1-175, and in *Report of the Walkerton Inquiry, Part One (WI, One)* (Toronto: Queen's Printer for Ontario, 2002), 7-12. http://mail.tscript.com/trans/walkerton2000.htm.
4 *Globe and Mail,* 26 May 2000, A9; 27 May 2000, A12, A19; 27 May 2000, 1.
5 http://mail.tscript.com/trans/walkerton2000.htm. Janice Hallahan questioned by Freyja Krisjanson, 15 November 2000; Jim Kieffer questioned by John Grace and Paul Cavalluzzo, 16 November 2000; Stan Koebel questioned by Glenn Hainey, 20 December 2000.
6 *Toronto Star,* 30 May 2000, A7.
7 *Toronto Star,* 1 May 2000, A6; *Globe and Mail,* 29 May 2000, A10; *Globe and Mail,* 31 May 2000, A1, A8; *Walkerton Herald-Times,* 7 June 2000, 1.
8 "Water Tragedy Is a Sign of Rural Municipal Incompetence," *Toronto Star,* 31 May 2000, A31.
9 *Globe and Mail,* 9 June 2000, A15; *Globe and Mail,* 20 June 2000, A6; *Globe and Mail,* 10 June 2000, A14.
10 "Community Has Strength to Overcome Tragedy and 'Town in Deep Sorrow,'" *Walkerton Herald-Times,* 31 May 2000, 4; "Under My Hat," *Walkerton Herald-Times,* 7 June 2000, 5; Letter to the Editor, "Recent Disaster Opens People's Eyes to the True Value of Town Officials," *Walkerton Herald-Times,* 14 June 2000, 8.
11 "Media Feeding Frenzy ..." *Walkerton Herald-Times,* 31 May 2000, 5; "Eerie Silence Fills Air," *Walkerton Herald-Times,* 7 June 2000, 4.
12 Bruno Latour, *We Have Never Been Modern* (Cambridge, MA: Harvard University Press, 1993); Donna Haraway, *Simians, Cyborgs and Women: The Reinvention of Nature* (New York: Routledge, 1991); Erik Swyngedouw, "Modernity and Hybridity: Nature, *Regeneracionismo,* and the Production of the Spanish Waterscape, 1890-1930," *Annals of the Association of American Geographers* 89, 3 (1999): 91-102, sec. 2; Swyngedouw, "On Hybrids and Socio-Nature: Flow, Process and Dialectics," in *El Problema des Abastecimiento de Aqua Potable en Equador* (Quito, Ecuador: ILSIS, 1995).
13 *Report of the Walkerton Inquiry, Part Two (WI, Two),* 279, 282. http://mail.tscript.com/trans/walkerton2000.htm; A. Sancton and T. Janik, "Provincial-Local Relations and the Drinking Water in Ontario," *Walkerton Inquiry Commissioned Paper 3* (Toronto: Ministry of the Attorney General, Walkerton Inquiry, 2002), 30-31.
14 *WI, Two,* 279-80. The province developed the capacity to deliver water services directly in 1956, through the Ontario Water Resources Commission: "The recent trend has been for the delivery of water services to shift further back to the municipalities." In 2001, 70 percent of municipal water systems were operated by the municipality. See *WI, Two,* Jim Boulden questioned by Frank Morocco, 29 November 2000.

15 Jamie Benidickson, "Water Supply and Sewage Infrastructure in Ontario, 1880-1990s," *Walkerton Inquiry Commissioned Paper 1*. All the commissioned papers are available on the CD version of the *Report of the Walkerton Inquiry* (Toronto: Queen's Printer for Ontario, 2002).

16 *WI, Two*, Frank Koebel questioned by Brian Gover, 6 December 2000; by Mike Epstein, 7 December 2000; and by Paul Burnstein, 7 December 2000.

17 *WI, One*, 183.

18 *Walkerton Herald-Times*, 21 June 2000, 6.

19 Ibid.; Ibid., 28 June 2000, 4, 5.

20 *WI, Two*, Jim Boulden questioned by Paul Cavalluzzo, 28 November 2000; by William Trudell, 29 November 2000; Stan Koebel questioned by Brian Gover, 19 December 2000.

21 Perkel, *Well of Lies*, 209.

22 The debate about chlorination, disinfection by-products, and cancer risks is also, of course, being carried on among experts, although I have no sign that town residents were directly aware of this controversy. See Joe Thornton, *Pandora's Poison: Chlorine, Health and a New Environmental Strategy* (Cambridge, MA: MIT Press, 2000), chap. 5 on organochlorines and cancer. The standard authority on chlorination is George Clifford White, *Handbook of Chlorination and Alternative Disinfectants*, multiple editions. See the refutation of associations between chlorine and cancer in the fourth edition (New York: Wiley, 1999), xv-xx and chap. 12.

23 *WI, Two*, Jim Boulden questioned by Paul Cavalluzzo; Frank Koebel questioned by Brian Gover, 6 December 2000; *WI, One*, 206.

24 "Utility Head Pleading Ignorance," *Toronto Star*, 28 November 2000, A4; Joel Axler, "*E. Coli*, Paralysis and Complacency," *Globe and Mail*, 6 June 2000, A17; *WI, Two*, Jim Kieffer questioned by Paul Cavalluzzo; Janice Hallahan questioned by Freyja Kristjanson; Walkerton Public Library, *Commission Exhibits*, Exhibits prepared to accompany the testimony of Janice Hallahan, no. 105, tab 6, "Rate Comparisons, Walkerton, Hanover, Kincardine, Port Elgin, Wingham," 5; *WI, Two*, "The Role of Municipal Governments," 299, 302, 312, 314. The commissioner found that the overuse of reserves, as appeared to be the case in Walkerton, reflected an "unduly conservative approach to financing." The Walkerton Public Utilities Commission had about $347,000 in its reserve fund as of 1 January 2000.

25 *WI, One*, 16, 193; *WI, Two*, Frank Koebel questioned by Brian Gover, 6 December 2000; Stan Koebel questioned by Brian Gover, 20 December 2000; Stan Koebel questioned by William Trudell, 20 December 2000; Jim Kieffer questioned by Joseph DiLuca, 16 November 2000.

26 *Globe and Mail*, 15 June 2000, A1, A9; *Walkerton Herald-Times*, 21 June 2000, 1.

27 *WI, Two*, Phil Bye questioned by Paul Cavalluzzo, 13 November 2000, 50; Frank Koebel questioned by Brian Gover, 6 December 2000, 7 December 2000; Frank Koebel questioned by Paul Burnstein; *Globe and Mail*, 2 June 2000, A6, quoting former mayor Jim Boulden; Axler, "*E. Coli*, Paralysis and Complacency," A17.

28 John Finlay, "Media Feeding Frenzy Eye-Opening," *Walkerton Herald-Times*, 31 May 2000, 5; Carol Barclay, "Community Has Suffered Enough," *Walkerton Herald-Times*, 28 June 2000, 4, 5; Elaine Crilly, "Will Our Water Be Safe Again?" *Walkerton Herald-Times*, 21 June 2000, 6.

29 Ulrich Beck, *Risk Society: Towards a New Modernity*, trans. Mark Ritter (London: Sage, 1992 [1986]).

30 "The Anatomy of a Disaster," *Toronto Star,* 3 June 2000, A4, col. 4; "Water Problems Raise Many Questions," *Walkerton Herald-Times,* 7 June 2000, 5; Beck, *Risk Society,* chap. 2.

31 "Special Report, the Anatomy of a Preventable Tragedy," *Toronto Star,* 14 October 2000, secs. S, S1-8.

32 Picture of Rotarian Tim Mancell reading the *Star* report, *Walkerton Herald-Times,* 1 November 2000, 1.

33 Sue Ann Ellis, "Decades of Fumbling at Heart of Crisis," *Walkerton Herald-Times,* 25 October 2000, 3, 4; "Story Evoked Many Emotions among Walkerton Residents," *Walkerton Herald-Times,* 1 November 2000, 1, 2.

34 Bruce-Grey-Owen Sound Health Unit, "The Investigative Report on the Walkerton Outbreak of Waterborne Gastroenteritis, May-June 2000," (Owen Sound: Bruce-Grey-Owen Sound Health Unit, 2000), iii, 54; Bruce-Grey-Owen Sound Health Unit, "Can't Assume Groundwater Sources Secure – Public Health Unit's Investigative Report," press release, 10 October 2000; "Report Questions Safety of All Deep-Drilled Wells," *Walkerton Herald-Times,* 18 October 2000.

35 *WI, One,* 99, 100; personal observation by the author, Walkerton Community Centre, 10 October 2000; *Walkerton Herald-Times,* 18 October 2000, 1, 2; *Toronto Star,* 16 October 2000, A4.

36 *Toronto Star,* 17 October 2000, A4.

37 Mary Douglas and Aaron Wildavsky, *Risk and Culture: An Essay on the Selection of Technological and Environmental Dangers* (Berkeley and Los Angeles: University of California Press, 1982), 3-6.

38 Ontario, Ministry of the Environment, *Drinking Water Objectives* (Toronto: MOE, 1984), 1.

39 *WI, One,* 30, 133; Pat Halpin, "MOE Officer's Report Critical of PUC," *Walkerton Herald-Times,* 15 November 2000, 9.

40 *WI, One,* 18, 28, 185.

41 "MOE Supervisor Didn't Think Walkerton Warranted 'Forceful Action,'" *Walkerton Herald-Times,* 22 November 2000, 11, 27; *WI, One,* 218-19, 227-38.

42 *WI, Two,* Mary Robinson-Ramsay questioned by Frank Morocco and by Ramani Nadarajah, 1 December 2000.

43 Perkel, *Well of Lies,* 203.

44 A good example is *WI, Two,* Stan Koebel questioned by Earl Cherniak, 20 December 2000, 169.

45 *WI, One,* 4, 19, 42, 229. The economists who prepared a study for the commission on provincial-local relations and drinking water in Ontario similarly reasoned from a unitary epistemology of interest: "Provincial testing requirements that reinforced the obvious self-interest in testing by the municipalities made perfect sense." "It is never in the economic interests of a particular community to provide itself with impure water although it might well be in the interests of the majority not to provide decent public transit or welfare systems." See A. Sancton and T. Janik, "Provincial-Local Relations and the Drinking Water in Ontario," *Walkerton Inquiry Commissioned Paper 3,* 3, 4, 35, 50.

46 For an analysis of how such certifications form knowledge, see Lawrence Busch and Keiko Tanaka, "Rites of Passage: Constructing Quality in a Commodity Subsector," *Science, Technology and Human Values* 21, 1 (1996): 3-27.

47 Benjamin Sims, "Concrete Practices: Testing in an Earthquake-Engineering Laboratory," *Social Studies* 29, 4 (1999): 488-89.

48 *WI, One*, 18, 28, 185.

49 Sancton and Janik, "Provincial-Local Relations," 5, 7, 8.

50 Justice O'Connor noted that the Koebels mistakenly believed the water supplying the Walkerton wells was safe, but he also noted that it was unclear "whether Stan and Frank Koebel would have altered their improper practices if they had received appropriate training." See *WI, One*, 28, 29. The reference to unguided practical experience is from Benidickson, "Water Supply," 132, citing B. Grover and D. Zussman, "Safeguarding Canadian Drinking Water," *Inquiry on Federal Water Resources, Research Paper No 4* (Ottawa: Inquiry on Federal Water Resources, 1985).

51 *Walkerton Herald-Times*, 11 October 2000, 6; Perkel, *Well of Lies*, 194.

52 John Finlay, "Compassion Goes Missing When Money on Line," *Walkerton Herald-Times*, 6 December 2000, 6.

53 "Cattlemen's Group Creates Trust Fund to Assist Biesenthals with Legal Bills," *Walkerton Herald-Times*, 6 December 2000, 2; Finlay, "Compassion," 6.

54 *Globe and Mail*, 12 December 2000, A1, A8; *Toronto Star*, 12 December 2000, A4; *Toronto Star*, 21 December 2000, A4; Pat Halpin, "PUC Records Faked Regularly," *Walkerton Herald-Times*, 13 December 2000, 1; Sue Ann Ellis, "Deal with Koebel Draws Criticism from Citizens," *Walkerton Herald-Times*, 13 December 2000, 1.

55 *Globe and Mail*, "Tragic Incompetence," 20 December 2000, A18; Joey Slinger, "Well of Self-Discipline Runs Deep at Utility," *Toronto Star*, 16 December 2000, A2; "Separated at Birth?" *Toronto Star*, 19 December 2000, A24.

56 *Toronto Star*, 18 December 2000, A1; *Globe and Mail*, 18 December 2000, A3.

57 *Walkerton Herald-Times*, 13 December 2000, 5; 20 December 2000, 5.

58 "Koebel's Severance Stymied by Council," *Toronto Star*, 9 January 2001, A17; "Water Plants Still Fail to Meet Standards, but This Doesn't Necessarily Equal Health Hazard, Government Says," *Toronto Star*, 22 December 2002; "Koebel Brothers to Face Charges, OPP Investigation Has Taken Nearly Three Years," *Toronto Star*, 22 April 2003; "OPP Is an Acronym for Ontario Provincial Police," *Walkerton Herald-Times*, 17 January 2001, 1; *Walkerton Herald-Times*, 4 September 2002, 1.

59 *Toronto Star*, 23 November 2002. The speaker quoted is Donnie Berberich, whose two youngest daughters were gravely ill during the crisis of spring 2000.

CHAPTER 8: CONCLUSION

1 Joy Parr, Jon van der Veen, and Jessica Van Horssen, "The Practicing of History Shared across Differences: Needs, Technologies and Ways of Knowing in the Megaprojects New Media Project," *Journal of Canadian Studies* 43, 1 (2009): 1-24.

2 See http://megaprojects.uwo.ca.

3 See http://megaprojects.uwo.ca/Gagetown; www.bgcha.ca/index.html; and www.rootsweb.ancestry.com/~nbbgcha.

4 "Turning Seaway Valley into Magnificent Park," *Ottawa Journal*, 24 September 1959.

5 Upper Canada Village Archives 9-12-1, Jeanne Minhinnick, "Expressions of Opinions of the Future Plans for Upper Canada Village"; file "Upper Canada Village, Principles Involved," Ronald L. Way, "Control of Upper Canada Village Restoration," 29 April 1958; Nicole V. Champeau, *Mémoire des villages engloutis: la voie Maritime du Saint-Laurent de Milles Roches aux Mille-Îles* (Ottawa: Vermillon, 1998), 19-20, 96-97.

6 See www.lostvillages.ca and "Francoise A. Laflamme," in "Lives Lived," *Globe and Mail*, 18 July 2000.

7 *The Columbia Kootenay Symposium*, videorecording, (Nelson, BC: Shaw Cable 10, 1993).

8 http://www.cbt.org.

9 http://www.ontarioparks.com/English/inve.html; and http://www.ontarioparks.com/English/macg.html.

10 There is an exploration of this connection in the popular study by Kim Todd, *Tinkering with Eden: A Natural History of Exotic Species in America* (New York: Norton, 2001).

11 Kim Collins-Lauber, "Nuclear Neighbour Speaks Out," *Whitecourt Star*, 9 July 2008.

12 *Owen Sound Sun-Times*, 15 July 2008.

Select Bibliography

Abraham, Itty. *The Making of the Indian Atomic Bomb.* London: Zed Books, 1998.

Adams, Paul C. "Peripatetic Imagery and the Peripatetic Sense of Place." In Paul C. Adams, Steven Hoelscher, and Karen E. Till, eds., *Textures of Place: Exploring Humanist Geographies.* Minneapolis: University of Minnesota Press, 2001, 186-206.

Adelson, Leslie. *Making Bodies, Making History: Feminism and German Identity.* Lincoln: University of Nebraska Press, 1993.

Allen, Barbara L. "Narrating the Toxic Landscape in 'Cancer Alley,' Louisiana." In David Nye, ed., *Technologies of Landscape: From Reaping to Recycling.* Boston: University of Massachusetts Press, 1999, 187-207.

Amato, Joseph. *On Foot: A History of Walking.* New York: New York University Press, 2004.

Armstrong, Austin, and Jennifer Walker. *Memories of Life in Two Counties.* Hampton, NB: privately published, 1998.

Beck, Ulrich. *Ecological Politics in an Age of Risk.* London: Polity, 1995.

–. *Enlightenment: Essays on the Politics of the Risk Society.* Atlantic Highlands, NJ: Humanities International Press, 1995.

–. *Risk Society: Towards a New Modernity.* Trans. Mark Ritter. London: Sage, 1986.

Benidickson, Jamie. "Water Supply and Sewage Infrastructure in Ontario, 1880-1990s." *Walkerton Inquiry Commissioned Paper 1.* Toronto: Ontario Ministry of the Attorney General, 2002.

Benjamin, Walter. *Illuminations.* New York: Harcourt Brace, 1968.

Betts, Matthew, and David Coon. *Working with the Woods: Restoring Forests and Community in New Brunswick.* Fredericton: Conservation Council of New Brunswick, 1996.

Blumm, Michael. *Sacrificing the Salmon: A Legal and Policy History of the Decline of Columbia Basin Salmon.* Den Bosch, The Netherlands: BookWorld Publishers, 2002.

Bocking, Stephen. *Nature's Experts: Science, Politics and the Environment.* New Brunswick, NJ: Rutgers University Press, 2004.

Bothwell, Robert. *Nucleus: The History of Atomic Energy of Canada Limited.* Toronto: University of Toronto Press, 1988.

Bouchard, Randy, and Dorothy Kennedy. *First Nations Ethnography and Ethnohistory in British Columbia's Lower Kootenay/Columbia Region.* Castlegar: Columbia Power Corporation, August 2000 (April 2005).

Bourdieu, Pierre. *The Logic of Practice.* Cambridge: Polity, 1990.

–. *Outline of a Theory of Practice.* New York: Cambridge University Press, 1977.

Bowser, Sara. "The Planners' Part." *Canadian Architect*, February 1958, 39.

Bremmer, Jan, and Herman Roodenburg. *A Cultural History of Gesture.* Cambridge: Polity, 1991.

Bremner, Dawn, ed. *Places of Our Hearts: Memories of Base Gagetown Communities to 1953.* Gagetown: Base Gagetown Community History Association and Queens County Historical Society and Museum, 2003.

Brown, Phil, and Faith T. Ferguson. "'Making a Big Stink': Women's Work, Women's Relationships and Toxic Waste Activism." *Gender and Society* 9 (1995): 145-72.

Busch, Lawrence, and Keiko Tanaka. "Rites of Passage: Constructing Quality in a Commodity Subsector." *Science, Technology and Human Values* 21, 1 (1996): 3-27.

Bush, Catherine. *The Rules of Engagement.* Toronto: HarperCollins, 2000.

Butler, Judith. *Bodies That Matter: On the Discursive Limits of Sex.* New York: Routledge, 1993.

–. *Gender Trouble: Feminism and the Subversion of Identity.* New York: Routledge, 1990.

Canning, Kathleen. "The Body as Method? Reflections on the Place of the Body in Gender History." In her *Gender and History in Practice: Historical Perspectives on Bodies, Class and Citizenship*, 168-190. Ithaca: Cornell University Press, 2005.

Carlisle, Rodney P. "Probabilistic Risk Assessment in Nuclear Reactor." *Technology and Culture* 38, 4 (1997): 920-41.

Carson, Rachel. *Silent Spring.* Greenwich, CT: Fawcett, 1962.

Casey, Edward S. *Getting Back into Place: Toward a Renewed Understanding of the Place-World.* Bloomington: Indiana University Press, 1993.

–. "How to Get from Space to Place in a Fairly Short Stretch of Time." In Steven Feld and Keith Basso, eds., *Senses of Place.* Santa Fe, NM: School of American Research Press, 1996, 13-52.

–. *Remembering: A Phenomenological Study.* Bloomington: Indiana University Press, 1987.

Champeau, Nicole V. *Mémoires des villages engloutis : La voie maritime du Saint-Laurent de Milles Roches aux Mille-Îles.* Ottawa: Vermillon, 1999.

Chiang, Connie Y. "Monterey-by-the-Smell: Odors and Social Conflict on the California Coastline." *Pacific Historical Review* 73, 2 (2004): 183-214.

Classen, Constance, David Howes, and Anthony Synnott. *Aroma, the Cultural History of Smell.* New York: Routledge, 1994.

Coates, Peter. "The Strange Stillness of the Past: Toward an Environmental History of Sound and Noise." *Environmental History* 10, 4 (2005): 636-65.

Connerton, Paul. *How Societies Remember.* New York: Cambridge University Press, 1989.

Corbett Richard, ed. *Summer Hill and Dunn's Corner: Gone But Not Forgotten.* Saint John, NB: Base Gagetown Community History Association, 2003.

Corbin, Alain. *The Foul and the Fragrant: Odor and the French Social Imagination.* Cambridge, MA: Harvard, 1986.

–. *The Life of an Unknown: The Rediscovered World of a Clogmaker in Nineteenth-Century France*. New York: Columbia University Press, 2001.

–. *The Lure of the Sea: The Discovery of the Seaside in the Western World, 1750-1840*. Oxford: Blackwell, 1994.

–. *Time, Desire and Horror: Towards a History of the Senses*. Cambridge, MA: Polity Press, 1995.

–. *Village Bells: Sound and Meaning in the Nineteenth-Century French Countryside*. New York: Columbia University Press, 1998.

Corn, Joe. "'Textualizing Technics': Owner's Manuals and the Reading of Objects." In Ann Smart Martin and J. Ritchie Garrison, eds., *American Material Culture: The Shape of the Field*. Winterthur, DE: Winterthur Museum, 1997, 169-94.

Coronil, Fernando. "Smelling Like a Market." *American Historical Review* 106 (February 2001): 119-29.

Cowan, Robin. "Nuclear Power: A Study in Technological Lock-in." *Journal of Economic History* 50, 3 (1990): 541-67.

Cowan, Ruth Schwartz. *More Work for Mother: The Ironies of Household Technology from the Open Hearth to the Microwave*. New York: Basic Books, 1983.

Cronon, William. "A Place for Stories: Nature, History and Narrative." *Journal of American History* 78 (March 1992): 1347-76.

–. "The Trouble with Wilderness: Or, Getting Back to the Wrong Nature." In Char Miller and Hal Rothman, eds., *Out of the Woods: Essays in Environmental History*. Pittsburgh: University of Pittsburgh, 1997, 28-50.

Csikszentmihalyi, Mihaly. *Flow: The Psychology of Optimal Experience*. New York: Harper and Row, 1990.

Csordas, Thomas J. "Modes of Somatic Attention." *Cultural Anthropology* 8 (1993): 135-56.

de Certeau, Michel. *The Practice of Everyday Life*. Berkeley: University of California Press, 1984.

Denby, Connie, and Carol Lawson, eds. *Along Hibernia Roads: Faded Memories*. Gagetown: Base Gagetown Community History Association, 2003.

Descartes, René. *Philosophical Writings of René Descartes*. Vol. 2: *Meditations on First Philosophy*. Trans. J. Cottingham, R. Stoothoff, and D. Murdoch. Cambridge: Cambridge University Press, 1984 [1641]).

Donahue, Brian. *The Great Meadow: Farmers and the Land in Colonial Concord*. New Haven: Yale University Press, 2004.

Douglas, Mary. *Purity and Danger: An Analysis of Concepts of Pollution and Taboo*. New York: Praeger, 1966.

Douglas, Mary, and Aaron Wildavsky. *Risk and Culture: An Essay on the Selection of Technological and Environmental Dangers*. Berkeley and Los Angeles: University of California Press, 1982.

Dubord, Larry E. *The Happy Hookers*. Privately published, 1988.

Duden, Barbara. *Women beneath the Skin: A Doctor's Patients in Eighteenth-Century Germany*. Cambridge, MA: Harvard University Press, 1991.

Duden, Barbara. *Disembodying Women: Perspectives on Pregnancy and the Unborn*. Cambridge, MA: Harvard University Press, 1993.

Dunlap, Thomas R. "Australian Nature, European Culture: Anglo Settlers in Australia." In Char Miller and Hal Rothman, eds., *Out of the Woods: Essays in Environmental History*. Pittsburgh: University of Pittsburgh, 1997, 273-89.

Dyke, D. "Relocation of 95 New Brunswick Farm Families, 1956-60." *Economic Annalist,* October 1962, 4-11.

Dyke, D., and F. Lawrence. "Relocation Adjustments of Farm Families." *Economic Annalist,* February 1960, 101-10.

Dynesius, Mats, and Christer Nillsson. "Fragmentation and Flow Regulation of River Systems in the Northern Third of the World." *Science,* 4 November 1994, 4.

Erickson, Kai. "Radiation's Lingering Dread." *Bulletin of the Atomic Scientists* 47 (March 1991): 34-39.

Ericsson, Fredrik, and Anders Avdic. "Information Technology and Knowledge Acquisition in Manufacturing Companies: A Scandinavian Perspective." In Elayne Coakes, Dianne Willis, and Steve Clarke, eds., *Knowledge Management in the Socio-Technical World.* Berlin: Springer, 2002, 121-35.

Evenden, Matthew D. *Fish versus Power: An Environmental History of the Fraser River.* Cambridge: Cambridge University Press, 2004.

Fahrni, Magda. *Household Politics: Montreal Families and Postwar Reconstruction.* Toronto: University of Toronto Press, 2005.

Faraghar, John Mack. *A Great and Noble Scheme: The Tragic Story of the Expulsion of the French Acadians from their American Homeland.* New York: Norton, 2005.

Feld, Steven. "Places Sensed, Senses Placed: Toward a Sensuous Epistemology of Environments." In David Howes, ed., *Empire of the Senses: The Sensual Culture Reader.* Oxford: Berg, 2005, 181-88.

Fiege, Mark. *Irrigated Eden: The Making of an Agricultural Landscape in the American West.* Seattle: University of Washington Press, 1999.

Finkel, Alvin. *Our Lives: Canada after 1945.* Toronto: Lorimer, 1997.

Fisher, Frank. *Citizens, Experts and the Environment: The Politics of Local Knowledge.* Durham, NC: Duke University Press, 2000.

Forbes, E.R., and D.A. Muise. *The Atlantic Provinces in Confederation.* Toronto: University of Toronto Press, 1993.

French, Maida Parlow. *Apples Don't Just Grow.* Toronto: McClelland and Stewart, 1954.

Gilles, R. Peter, and Thomas R. Roach. *Lost Initiatives: Canada's Forest Industries: Forest Policy and Conservation.* New York: Greenwood, 1986.

Gillette, Robert. "'Transient' Nuclear Workers: A Special Case for Standards." *Science* 186, 11 October 1974, 125-29.

Gobba, Fabriziomaria. "Occupational Exposures to Chemicals and Sensory Organs: A Neglected Field of Study." *Neurotoxicology* 24 (August 2003): 675-91.

Godfrey, Sima. "Alain Corbin: Making Sense of French History." *French Historical Studies* 25 (2002): 381-98.

Goldsmith Edward, and Nicholas Hildyard. *The Social and Environmental Effects of Large Dams.* Camelford, Cornwall: Wadebridge Ecological Centre, 1984.

Gowing, Laura. *Women, Touch and Power in Seventeenth-Century England.* New Haven: Yale University Press, 2003.

Gray, John. "Open Spaces and Dwelling Places: Being at Home on Hill Farms in the Scottish Borders." *American Ethnologist* 26 (1999): 440-60.

Grosz, Elizabeth. "Bodies and Knowledges: Feminism and the Crisis of Reason" In *Space, Time and Perversion: Essays on the Politics of Bodies.* 25-44 London and New York: Routledge, 1995.

Halenbeck, William H. *Radiation Protection.* Boca Raton: Lewis, 1994.

Hales, Peter. *Atomic Spaces: Living on the Manhattan Project.* Urbana: University of Illinois, 1997.

Hall, Edward T. *The Hidden Dimension.* Garden City, NY: Doubleday, 1966.

–. *The Silent Language.* Garden City, NY: Doubleday, 1959.

Hansen, Mark. *Embodying Technesis: Technology beyond Writing.* Ann Arbor: University of Michigan Press, 2000.

Haraway, Donna. *Simians, Cyborgs and Women: The Reinvention of Nature.* London: Free Association Books, 1991.

Harden, Blaine. *A River Lost: The Life and Death of the Columbia.* New York: Norton, 1996.

Hardy, René, and Normand Séguin. *Forêt et Société en Mauricie.* Québec: Laval, 2004.

Harper, Douglas. *Changing Works: Visions of a Lost Agriculture.* Chicago: University of Chicago Press, 2001.

Harper, Douglas. *Working Knowledge: Skill and Community in a Small Shop.* Chicago: University of Chicago Press, 1987.

Harvey, David. *The Condition of Postmodernity: An Inquiry into the Origins of Cultural Change.* Oxford: Blackwell, 1989.

Hayes, Samuel P. *Beauty, Health and Permanence: Environmental Politics in the United States.* New York: Cambridge University Press, 1987.

Hayles, N. Katherine. "Foreword: Clearing the Ground." In Mark Hansen, *Embodying Technesis: Technology beyond Writing.* Ann Arbor: University of Michigan Press, 2000.

–. "Searching for Common Ground." In Michael E. Soulé and Gary Lease, eds., *Reinventing Nature? Responses to Postmodern Deconstruction.* Washington, DC: Island Press, 1995, 46-63.

–. "Situated Nature and Natural Simulations: Rethinking the Relation between the Beholder and the World." In William Cronon, ed., *Uncommon Ground: Toward Reinventing Nature.* New York: Norton, 1995, 329-78.

Hazlett, Maril. "'Woman vs. man vs. bugs': Gender and Popular Ecology in Early Reactions to Silent Spring." *Environmental History* 9 (2004): 701-29.

Hecht, Gabrielle. *The Radiance of France: Nuclear Power and National Identity after World War II.* Cambridge, MA: MIT Press, 1998.

Herron John P., and Andrew G. Kirk, eds. *Human/Nature: Biology, Culture, and Environmental History.* Albuquerque: University of New Mexico Press, 1999.

Hills, T.L. *The St. Lawrence Seaway.* London: Methuen, 1959.

Hirt, Paul. *Terra Pacifica: People and Place in the Northwest States and Western Canada.* Pullman: Washington State University Press, 1998.

Hornig, James. *Social and Environmental Impact of the James Bay Hydroelectric Project.* Montreal: McGill-Queen's University Press, 1999.

Hull, Isabel V. "The Body as Historical Experience: Review of Recent Works by Barbara Duden." *Central European History* 28 (1995): 75-79.

Jackson, J.H. "The St. Lawrence Power Project Rehabilitation: A Review of Major Factors." *Engineering Journal* (February 1960).

Jasanoff, Sheila. "Civilisation and Madness: The great BSE Scare of 1996." *Public Understanding of Science* 6, 3 (1997): 221-32.

Jenner, Mark S.R. "Civilisation and Deodorisation? Smell in Early Modern English Culture." In Peter Burke, Brian Harrison, and Paul Slack, eds., *Civil Histories: Essays Presented to Sir Keith Thomas.* Oxford: Oxford University Press, 2000, 127-44.

Jermyn, Chris. "Some St. Lawrence Seaway Communities, 1959-1969." *Canadian Geographical Journal* 79 (1969): 155-63.

Johnson, Franklin. "Community of New Jerusalem." In Verna Mott, ed., *New Jerusalem We Remember*. Gagetown: Base Gagetown History Association, 2003.

Keller, Evelyn Fox. *A Feeling for the Organism: The Life and Work of Barbara McClintock*. San Francisco: W.H. Freeman, 1983.

Killan, Gerald. *Protected Places: A History of Ontario's Provincial Parks System*. Toronto: Dundurn Press, 1995.

Krutilla, John. *The Columbia Basin Treaty: The Economics of an International River Basin*. Baltimore: Johns Hopkins University Press, 1967.

Latour, Bruno. *We Have Never Been Modern*. Cambridge, MA: Harvard University Press, 1993.

Lawless, H.T., and T. Engen. "Associations to Odors: Inference, Mnemonics and Verbal Labelling." *Journal of Experimental Psychology* 3, 1 (1977): 52-59.

Laxer, James. *The Acadians in Search of a Homeland*. Toronto: Doubleday, 2006.

Lefebvre, Henri. *The Production of Space*. Oxford: Blackwell, 1991.

Leiss, William. "Between Expertise and Bureaucracy: Risk Management Trapped at the Science-Policy Interface." In G. Bruce Doern and Ted Reed, eds., *Risky Business: Canada's Changing Science-based Policy and Regulatory Regime*. Toronto: University of Toronto Press, 2000, 49-74.

Loo, Tina. "People in the Way: Modernity, Environment, and Society on the Arrow Lakes." *BC Studies* 142 (Summer/Autumn 2004): 43-77.

Lower, A.R.M. *Settlement and the Forest Frontier in Eastern Canada*. Toronto and New Haven: Ryerson and Yale University Press, 1938.

Marin, Clive, and Frances Marin. *Stormont, Dundas and Glengarry, 1945-78*. Belleville: Mika, 1982.

Martin, Angela K. "The Practice of Identity and an Irish Sense of Place." *Gender, Place and Culture* 4, 1 (1997): 89-114.

Marx, Karl. "Private Property and Communism." In D.J. Struick, ed., *The Economic and Philosophical Manuscripts of 1844*. New York: International Publishers, 1972.

Maturana, Humberto, and Bernhard Poerksen. *From Being to Doing: The Origins of the Biology of Cognition*. Heidelberg: Carl-Auer, 2004.

Maturana, Humberto, and Francisco Varela. *Tree of Knowledge: The Biological Roots of Human Understanding*. Boston: Shambhala, 1992.

Mauss, Marcel. "Les techniques du corps." *Journal de psychologie (Sociologie et anthropologie)* 1936: 363-86.

Mabee, Carleton. *The Seaway Story*. New York: Macmillan, 1961.

Mazur, Allan. *Hazardous Inquiry: The Rashomon Effect at Love Canal*. Cambridge, MA: Harvard University Press, 1998.

–. *True Warnings and False Alarms: Evaluating Fears About the Health Risks of Technology, 1948-1971*. Washington, DC: Resources for the Future, 2004.

McCully, Patrick. *Silenced Rivers: The Ecology and Politics of Large Dams*. London: Zed Books, 1996.

McLean, Stuart. *Welcome Home: Travels in Small Town Canada*. Toronto: Viking, 1992.

McLuhan, Marshall. *Gutenberg Galaxy: The Making of Typographic Man*. Toronto: University of Toronto Press, 1962.

–. *Understanding Media: The Extensions of Man*. New York: McGraw-Hill, 1964.

Merchant, Carolyn. *Ecological Revolutions: Nature, Gender and Science in New England.* Chapel Hill: University of North Carolina Press, 1989.

–. "Gender and Environmental History." *Journal of American History* 76 (1990): 1117-21.

–. "Reinventing Eden: Western Culture as a Recovery Narrative." In William Cronon, ed., *Uncommon Ground: Toward Reinventing Nature.* New York: Norton, 1995, 132-70.

Merleau-Ponty, Maurice. *Phenomenology of Perception.* New York: Humanities Press, 1962.

Miller, Char, and Hal Rothman, eds. *Out of the Woods: Essays in Environmental History.* Pittsburgh: University of Pittsburgh, 1997.

Mitchell, Lisa M., and Alberto Cambrosio. "The Invisible Topography of Power: Electromagnetic Fields, Bodies and the Environment." *Social Studies of Science* 27, 2 (1997): 221-71.

Mitman, Gregg. *Breathing Space: How Allergies Shape Our Lives and Landscapes.* New Haven: Yale University Press, 2007.

Moen, I.O. *Investigation of the Explosion Hazards of Hydrogen Sulphide, Phase II.* Ottawa: AECB, 1986. Toronto: Micromedia, 1994 microlog 88-06131 1, 2, 6.

Mohai, Paul. "Men, Women and the Environment: An Examination of the Gender Gap in Environmental Concern and Activism." *Society and Natural Resources* 5 (1992): 1-19.

Mott, Verna, ed. *New Jerusalem: We Remember.* Gagetown: Base Gagetown History Association, 2003.

Mumford, Lewis. "The Highway and the City." *Architectural Record* 123 (April 1958): 179-86.

Murphy, Michele. "The 'Elsewhere within Here,' and Environmental Illness: Or, How to Build Yourself a Body in a Safe Space." *Configurations* 8, 1 (2000): 87-120.

Nash, Linda. "Finishing Nature: Harmonizing Bodies and Environments in Late-Nineteenth-Century California." *Environmental History* 8 (January 2003): 25-52.

–. *Inescapable Ecologies: A History of Environment, Disease and Knowledge.* Berkeley: University of California Press, 2006.

New Brunswick Power. *Radiation Protection Training Course,* 1979, 1981, 1986, 1992, 2002.

New Brunswick, Royal Commission on the Milk Industry in New Brunswick, *Report,* 1970-71. Fredricton: Government of New Brunswick.

Newman, Bo. "Agents, Artifacts and Transformations: The Foundations of Knowledge Flows." In Clyde W. Holsapple, ed., *Handbook on Knowledge Management I.* Berlin: Springer, 2003.

Newman, Simon. *Embodied History: The Lives of the Poor in Early Philadelphia.* Philadelphia: University of Pennsylvania Press, 2003.

Nikiforuk, Andrew. *Saboteurs: Wiebo Ludwig's War against Big Oil.* Toronto: Macfarlane Walter and Ross, 2002.

Norwood, Vera. "Constructing Gender." In J.P. Herron and A.G. Kirk, eds., *Human Nature: Biology, Culture and Environmental History.* Albuquerque, NM: University of New Mexico, 1999, 49-62.

Nuclear Energy Agency, OECD. *Work Management in the Nuclear Power Industry.* Paris: OECD, 1997.

Nye, David. "Technology, Nature and American Origin Stories." *Environmental History* 8 (2003): 8-24.

Oelschlaeger, L. Max. "Re-Placing History, Naturalizing Culture." In J.P. Herron and A.G. Kirk, eds., *Human Nature: Biology, Culture and Environmental History.* Albuquerque: University of New Mexico, 1999, 63-78.

Ong, Walter J. "The Shifting Sensorium." In *The Presence of the Word.* New Haven: Yale, 1967, 1-9.

Ontario, Ministry of the Environment. *Drinking Water Objectives.* Toronto: MOE, 1984.

Parent, Milton. *Port of Nakusp.* Nakusp: Arrow Lakes Historical Society, 1992.

Parenteau, William. "Forest and Society in New Brunswick: The Political Economy of Forest Industries." PhD. diss., University of New Brunswick, 1994.

–. "'In Good Faith': The Development of Pulpwood Marketing for Independent Producers in New Brunswick, 1960-1975." In Anders L. Sandberg, ed., *Trouble in the Woods: Forest Policy and Social Conflict in Nova Scotia and New Brunswick.* Fredericton, NB: Acadiensis Press, 1992.

Parr, Joy. *Domestic Goods: The Material, the Moral and the Economic in the Postwar Years.* Toronto: University of Toronto Press, 1999.

–. "Local Water Diversely Known: Walkerton 2000 and After." *Environment and Planning D: Society and Space* 23, 2 (2005): 251-71.

–. "Smells Like? Sources of Uncertainty in the History of a Great Lakes Environment." *Environmental History* 11 (2006): 282-312.

Paskievici, Wladimir, and L. Zikovsky. "Public Health Risks Associated with the CANDU Nuclear Fuel Cycle: Non-Radiological Risks." Ottawa: Atomic Energy Control Board, September 1982, 70. Toronto: Micromedia, 1994 microlog 85-02818.

Pearkes, Eileen Delehanty. *The Geography of Memory: Recovering Stories of a Landscape's First People.* Nelson: Kutenai House Press, 2002.

Perkel, Colin. *Well of Lies: The Walkerton Tragedy.* Toronto: McClelland and Stewart, 2002.

Petersen, Keith. *River of Life, Channel of Death: Fish and Dams on the Lower Snake.* Corvallis: Oregon State University Press, 1995.

Petrie, Joseph Richards. *The Regional Economy of New Brunswick.* Fredericton: Canadian Committee on Reconstruction Report, 1943.

Polanyi, Michael. *The Tacit Dimension.* New York: Anchor Books, 1967.

Porter, John. *The Vertical Mosaic: An Analysis of Social Class and Power in Canada.* Toronto: University of Toronto Press, 1965.

Porteus, J. Douglas. "Smellscape." *Progress in Geography* 9, 3 (1985): 356-78.

Power, Thomas Michael. *Lost Landscapes and Failed Economies.* Washington, DC: Island Press, 1996.

Prior, Michael, Michelle Mostrom, Robert Coppock, and Florence Zack. *Environmental Health Scoping Study at Bruce Heavy Water Plant,* AECB Project 3.168.1. Ottawa: AECB, October 1995.

Pryce, Paula. *"Keeping the Lakes' Way": Reburial and the Re-Creation of a Moral World among an Invisible People.* Toronto: University of Toronto Press, 1999.

Rae, H.K. *Canada Enters the Nuclear Age: A Technical History of Atomic Energy of Canada Limited.* Kingston and Montreal: McGill-Queen's University Press, 1997.

–. "Heat Transport System." In Atomic Energy Canada Limited, *Canada Enters the Nuclear Age: A Technical History of Atomic Energy of Canada Limited.* Kingston and Montreal: McGill-Queen's University Press, 1997, 283-87.

Reicker, Marion. *Those Days Are Gone Away: Queens County, N.B., 1643-1901.* Gagetown: Queens County Historical Society, 1981.

Reid, E.P., and J. Fitzpatrick. *Atlantic Provinces Agriculture.* Ottawa: Canada Department of Agriculture, Economics Division, August 1957.

Reiser, Stanley J. "Technology and the Use of the Senses in Twentieth-Century Medicine." In W.F. Bynum and Roy Porter, eds., *Medicine and the Five Senses.* New York: Cambridge University Press, 1993, 262-73.

Roach, Thomas. "Farm Woodlots and Pulpwood Exports From Eastern Canada, 1900-1930." In Harold K. Steen, ed., *History of Sustained-Yield Forestry: A Symposium.* Portland, OR: Forest History Society for the International Union of Forestry Research Organizations, 1984.

Rodaway, Paul. *Sensuous Geographies: Body, Sense, and Place.* New York: Routledge, 1994.

Rome, Adam. "From the Editor." *Environmental History* 10 (2005): 635.

Roy, Arundhati. *The Cost of Living.* New York: Viking, 1999.

Russell, Edmund. "Evolutionary History: Prospectus for a New Field." *Environmental History* 8 (April 2003): 204-28.

Saber, A.J., A. Sulmistras, and I.O. Moen. *Investigation of the Explosion Hazards of Hydrogen Sulphide, Phase I.* Ottawa: AECB, 1986. Toronto: Micromedia, 1994, microlog 86-05414 1, 3, 4, 7.

Sancton, A., and T. Janik. "Provincial-Local Relations and the Drinking Water in Ontario." *Walkerton Inquiry Commissioned Paper 3.* Toronto: Ministry of the Attorney General, Walkerton Inquiry, 2002.

Sandberg, L. Anders. *Trouble in the Woods: Forest Policy and Social Conflict in Nova Scotia and New Brunswick.* Fredericton: Acadiensis Press, 1992.

Sangster, Joan. *Dreams of Equality: Women on the Canadian Left, 1920-1950.* Toronto: McClelland and Stewart, 1989.

Scarry, Elaine. *The Body in Pain: The Making and Unmaking of the World.* Oxford: Oxford University Press, 1985.

–. *Resisting Representation.* Oxford: Oxford University Press, 1994.

Schafer, R. Murray. *The Tuning of the World.* New York: Knopf, 1977.

Scharff, Virginia. "Lighting Out for the Territory: Women, Mobility and Western Place." In Richard White and John Findlay, eds., *Place and Power in the North American West.* Seattle: University of Washington Press, 1999, 287-303.

–. "Of Parking Spaces and Women's Places: The Los Angeles Parking Ban of 1920." *National Women's Studies Association Journal* 1 (1988): 37-51.

–. *Taking the Wheel: Women and the Coming of the Motor Age.* New York: Free Press, 1991.

Scharmer, Claus Otto. "Self-Transcending Knowledge." In Ikujiro Nonaka and David Teese, eds., *Managing Industrial Knowledge.* London: Sage, 2001, 68-90.

Schoch-Spana, Monica. "Reactor Control and Environmental Management: A Cultural Account of Agency in the US Nuclear Weapons Complex." PhD. diss., Johns Hopkins University, 1998.

Schor, Juliet B. *The Overworked American: The Unexpected Decline of Leisure.* New York: Basic Books, 1991.

Scott, James. *Seeing Like a State: How Certain Schemes to Improve the Human Condition Have Failed.* New Haven: Yale University Press, 1998.

Scott, Joan. "The Evidence of Experience." *Critical Inquiry* 17 (1991): 773-97.

–. *Gender and the Politics of History.* New York: Columbia University Press, 1988.

–. "Gender: A Useful Category of Historical Analysis." *American Historical Review* 91 (1986): 1053-75.

See, Scott. *Riots in New Brunswick: Orange Nativism and Social Violence in the 1840s.* Toronto: University of Toronto Press, 1993.

Seeley, John, R., Alexander Sim, and Elizabeth W. Loosley. *Crestwood Heights: A Study of the Culture of Suburban Life.* Toronto: University of Toronto Press, 1956.

Séguin, Normand. "L'économie agro-forestiére." *Revue de l'histoire de l'Amérique français* 29, 4 (1976): 559-65.

Sellers, Christopher. "Thoreau's Body: Towards an Embodied Environmental History." *Environmental History* 4, 4 (1999): 486-514.

Sheller, Mimi. "Automotive Emotions." *Theory, Culture and Society* 21, 4/5 (2004): 221-42.

Sims, Benjamin. "Concrete Practices: Testing in an Earthquake-Engineering Laboratory." *Social Studies of Science* 29, 4 (1999): 483-518.

Smilor, Raymond. "Personal Boundaries in the Urban Environment: The Legal Attack on Noise, 1865-1930." In Char Miller and Hal Rothman, eds., *Out of the Woods: Essays in Environmental History.* Pittsburgh: University of Pittsburgh Press, 1997, 181-93.

Smith, Dorothy. *The Everyday World as Problematic: A Feminist Sociology.* Boston: Northeastern University Press, 1987.

Smith, Mark M. "Making Sense of Social History." *Journal of Social History* 37 (Fall 2003): 165-86.

Solnit, Rebecca. *Wanderlust: A History of Walking.* New York: Viking, 2000.

Stokes, Peter. "St. Lawrence: A Criticism." *Canadian Architect,* February 1958, 46.

Strasser, Susan. *Never Done: A History of American Housework.* New York: Pantheon, 1982.

Swainson, Neil. *Conflict over the Columbia: The Canadian Background to an Historic Treaty.* Montreal: McGill-Queen's University Press, 1979.

Swyngedouw, Erik. "On Hybrids and Socio-Nature: Flow, Process and Dialectics." In *El Problema des Abastecimiento de Aqua Potable en Equador.* Quito, Ecuador: ILSIS, 1995.

–. "Modernity and Hybridity: Nature, *Regeneracionismo,* and the Production of the Spanish Waterscape, 1890-1930." *Annals of the Association of American Geographers* 89, 3 (1999): 91-102.

Tarr, Joel. *The Search for the Ultimate Sink: Urban Pollution in Historical Perspective.* Akron, OH: University of Akron Press, 1996.

Taylor, Joseph E. *Making Salmon: An Environmental History of the Northwest Fisheries Crisis.* Seattle: University of Washington Press, 1999.

Thompson, Evan. "The Mindful Body: Embodiment and Cognitive Science." In Michael O'Donovan-Anderson, ed., *The Incorporated Self: Interdisciplinary Perspectives on Embodiment.* Lanham, MD: Rowman and Littlefield, 1996, 127-44.

Thornton, Joe. *Pandora's Poison: Chlorine, Health and a New Environmental Strategy.* Cambridge, MA: MIT Press, 2000.

Thrift, Nigel. "Driving in the City." *Theory, Culture and Society* 21, 4/5 (2004): 41-59.

Tuan, Yi-fu. *Topophilia: A Study of Environmental Perception, Attitudes, and Values.* Englewood Cliffs: Prentice-Hall, 1974.

Turney-High, Harry Holbert. *Ethnography of the Kutenai.* Menasha, WI: Memoirs of the American Anthropological Association, No. 56, 1941.

University of Toronto, School of Social Work. *Round Table on Man and Industry, Community Survey Report, St. Lawrence Impact Area,* vol. 4, part 2, Town of Iroquois, 40, 39, 41.

Urry, John. "Sensing the City." In Dennis R. Judd and Susan Fainstein, eds., *The Tourist City.* New Haven: Yale University Press, 1999, 71-88.

Valenčius, Conevery Bolton. *The Health of the Country: How American Settlers Understood Themselves and Their Land*. New York: Basic Books, 2002.

Varela, Francisco, Evan Thompson, and Eleanor Rosch. *The Embodied Mind: Cognitive Science and Human Experience*. Cambridge, MA: MIT Press, 1991.

Walker, J. Samuel. *Permissible Dose: A History of Radiation Protection in the Twentieth Century*. Berkeley: University of California Press, 2000.

Waterfield, Donald. *Continental Waterboy*. Toronto: Clarke, Irwin, 1970.

–. *Land Grab: Oliver Buerge versus the Authority*. Toronto and Vancouver: Clarke, Irwin, 1973.

White, George Clifford. *Handbook of Chlorination and Alternative Disinfectants*. 4th ed. New York: Wiley, 1999.

White, Hayden. *The Content of the Form: Narrative Discourse and Historical Representation*. Baltimore: Johns Hopkins University Press, 1987.

White, Richard. *Organic Machine: The Remaking of the Columbia River*. New York: Hill and Wang, 1995.

Wilson, J.W. *People in the Way: The Human Aspects of the Columbia River Project*. Toronto: University of Toronto Press, 1973.

Wilson, J.W., and Maureen Conn. "On Uprooting and Rerooting: Reflections on the Columbia River Project." *BC Studies* 58 (1983): 40-54.

Wilson, Norman D. "The Rehabilitation of the St. Lawrence Communities." Canada, Advisory Committee on Reconstruction, April 1943, section 4. Ottawa.

Wilson, R., G.A. Vivian, W.J. Chase, G. Armitage, and L.J. Sennema. "Occupational Dose Reduction Experience in Ontario Hydro Nuclear Power Stations." *Nuclear Technology* 72 (1986): 243-44.

Worster, Donald. *Rivers of Empire: Water, Aridity, and the Growth of the American West*. New York: Oxford University Press, 1992.

–. "Two Faces West: The Development Myth in Canada and the US." In Paul Hirt, ed., *Terra Pacifica: People and Place in the Northwest States and Western Canada*. Pullman: Washington State University Press, 1998.

–. *The Wealth of Nature: Environmental History and the Ecological Imagination*. New York: Oxford University Press, 1993.

Wright, J.V. *Knechtel 1 Site, Bruce County, Ontario*. Ottawa: National Museum of Man, 1972.

Wright, Miriam. "Young Men and Technology: Government Attempts to Create a 'Modern' Fisheries Workforce in Newfoundland, 1949-1970." *Labour/Le travail* 42 (1998): 143-59.

Wynn, Graeme. "Deplorably Dark and Demoralised Lumberers?" *Journal of Forest History* 24 (1980): 168-87.

–. *Timber Colony: A Historical Geography of Early Nineteenth Century New Brunswick*. Toronto: University of Toronto Press, 1981.

Zonabend, Françoise. *The Nuclear Peninsula*. Trans. J.A. Underwood. Cambridge: Cambridge University Press, 1993.

Index

Printed and bound in Canada by Friesens

Set in Garamond by Artegraphica Design Co. Ltd.

Copy editor: Joanne Richardson

Proofreader: Stephanie VanderMeulen

Indexer: Lillian Ashworth